Heat Exchangers: Selection, Rating and Thermal Design

Heat Exchangers: Selection, Rating and Thermal Design

Editor

Hariom Sharma

scitus
academics

Heat Exchangers: Selection, Rating and Thermal Design

Edited by **Hariom Sharma**

ISBN: 978-1-68117-398-6

Library of Congress Control Number: 2015941588

SCITUS Academics LLC,
www.scitusacademics.com
Box No. 4766, 616 Corporate Way,
Suite 2, Valley Cottage,
NY 10989

Notice

Printed in United States of America on Acid Free Paper ∞

Contents

Preface

Heat exchangers are devices in which heat is transferred from one fluid to another fluid as a result of temperature difference. The heat exchanger presented in this book, in which water flows inside the tubes, but outside the tubes, air flows to enable the circulating water to cool the engine of a machine. In a wide range of engineering applications, heat exchangers are important, including chemical industries, processes, aircraft, automobiles and power plants, and heating, air conditioning and cooling.

This book provides an organized treatment of the many types of heat exchangers, focusing on rating, hydraulic and thermal design and selection. The fundamentals of thermal design for microscale heat exchangers and the improvement of heat transfer for applications in the design of heat exchangers with nanofluids are investigated through heat transfer. In addition, it examines single-phase forced convection correlations and flow friction factors for pumping power and heat transfer calculations in microchannel flows. In every industrial process, heat exchangers range from fossil and geothermal power generation to desalination and cooling, in general, as an important component of energy transfer and saving systems. They are also present in all types of vehicles, household appliances and electronics, cooling, heating and air conditioning in our daily lives. This book is intended to provide information for researchers, academics, postgraduate students and engineers working in the field of research and development in heat exchangers.

Chapter 1

Diverse Configurations of the Boiler Feed Pump Drive for the Ultra-supercritical 900-MW Steam Plant

Katarzyna Stępczyńska, Henryk Łukowicz, and Sławomir Dykas

Institute of Power Engineering and Turbomachinery, Silesian University of Technology, Konarskiego 18, Gliwice, 44-100, Poland.

ABSTRACT

Coal-based electric power generation remains the basic source of obtaining energy. With increasing pressure to reduce CO_2 emissions, improving power unit efficiency has become an issue of utmost significance. Surely, one of the possibilities to improve the efficiency of new power units is raising the steam parameters. With improved power plant efficiency, there is a lower demand for power of almost all auxiliary equipment except the boiler feed pump. The reason for this is that the power needed to drive the feed pump is an almost linear function of the steam pressure. This means that, even though the steam mass flow (and, consequently, the feed water mass flow) is reduced and the efficiency of feed pumps is improved, their power increases. For this reason, it is very important to find the optimum drive of the boiler feed pump. The main aim of the conducted analysis was to compare various drive options of the boiler feed pump for a conceptual ultra-supercritical 900-MW steam power unit. The following drive configurations of the boiler feed pump were presented and compared:

- A frequency-controlled electric motor
- A condensing turbine fed with steam extraction from the immediate-pressure turbine
- An extraction-backpressure turbine fed with steam from a cold reheat steam line with bleeds shifted from the low-pressure turbine
- A backpressure turbine fed with steam from a hot reheat steam line operating in parallel with the intermediate-pressure turbine
- An extraction-backpressure turbine fed with steam from a cold reheat steam line with bleeds shifted from the intermediate-pressure turbine (the master cycle idea).

The analysis of the operation of the 900-MW unit with various configurations of the feed pump drive was carried out for three load levels: for the nominal mass flow of live steam and for the partial mass flow of 75% and 50%.

BACKGROUND

The perspective of ever more stringent restrictions on emissions and of the rising coal consumption pose a challenge to the professional power engineering and necessitate the search for effective and economically acceptable methods of reducing CO_2 emissions. In the next years to come, condensing coal-fired power units will still be the pillar of the Polish electric power system. In the present situation, it is, therefore, necessary to achieve better and better efficiency of electric energy generation which will have a critical significance in meeting the emission-related requirements. Undoubtedly, raising the steam parameters is the main driving factor in the efforts to increase efficiency. Thus, the construction of new generation coal-fired units for ultra-supercritical parameters is a matter of the next few years to come. The efficiency of power generation in conventional coal-fired units has increased in the last 10 years by approximately 7% points. The power units which are now being built in the pulverized coal-fired (PF) technologies reach a capacity of the order of 1,000 MW. The pacemakers in the electric power generation net efficiency are now: Nordjyland in Denmark (47% for hard coal) and the BoA power unit in Niederaussem, Germany (45.5% for brown coal) [1].

Improving a power unit efficiency results in a reduction in fuel and steam consumption per unit which involves a reduction in the mass flow of burned fuel, feed water and other combustion products: slag and ash. The expansion of regenerative systems and the raising of feed water temperature additionally result in reducing the mass flow of the main condensate and of the mass flow of the water cooling the condenser. Generally, an effect of an increase in the steam parameters is a decreased demand for power to satisfy the needs of most facilities. However, there is a greater need in terms of the power with which the feed water is pumped. Since the power of the boiler feed pump is an almost linear function of the live steam pressure despite a reduction in the unit steam consumption and the feed water consumption, as well as an improvement in the efficiency of the pump itself, its share in the total demand for power to satisfy own needs increases substantially.

Boiler Feed Pump Drives Currently in Use

The feed water system is one of the critical components of a power unit. Its properties decide about many parameters of the thermal cycle but also about a failure-free and safe operation of the power plant. Therefore, the problem of the selection of the boiler feed pump drive is one of the most important issues in designing power units. Due to the requirements, each power unit should be equipped with at least two boiler feed pumps whose nominal capacity should ensure the maximum capacity of the once-through boiler. The value of the pumping effective pressure is included in the range of 1.1-1.4 of the steam pressure rating in the boiler. The efficiency of modern feed pumps exceeds 85%. The demand for power while pumping feed water is from 2% to 4.2% of the power generated by the unit [2]. Consequently, the driving motor of the feed pump is usually the principal receiver of electricity in the power plant.

In high power units, an extra low speed booster pump is used to prevent cavitation. The pump, which features an inlet designed specially to avoid cavitation, feeds water into the main pump, thus, protecting it entirely from this phenomenon. The booster pump may have a separate drive, or it can be driven by the main pump driving motor through a speed reduction gear [3].

To drive boiler feed pumps in power units with a capacity of up to approximately 250 MW, induction squirrel-cage motors and a fluid coupling to adjust rotation speed are used almost exclusively. However, as the power capacity of newly constructed units grew, the lack of appropriate electric motors (EM) became a problem. For this reason, for power units with a capacity exceeding 300 MW, steam drives were applied by means of a separate turbine incorporated into the thermal cycle. The steam turbine starts to drive the feed pump only after a sufficient amount of steam has become available. Due to that, it is necessary to use an additional start-up pump with a partial capacity of 30% to 50% with an electric drive. The driving turbine of the boiler feed pump can be incorporated into the thermal cycle in two ways [2]:

- As a condensing turbine (CT), fed with steam from the intermediate-pressure turbine (outlet steam is condensed in a separate condenser or it is introduced into the main condenser)
- As a backpressure turbine (BPT) or extraction-backpressure turbine (EBPT), fed with steam from the high-pressure turbine (outlet steam and steam bleeds, if any, are condensed in low-pressure regenerative heaters).

Because condensing turbines used to drive boiler feed pumps have a limited effect on the thermal cycle of the unit, they have found a much wider application in the professional power engineering. In Poland, this configuration was used in the units of the power plants in Bełchatów, Opole, P tnów and Łagisza.

Another possible option is the feed pump driven by the shaft of the main turbine. The pump can be joined to the turbine shaft either at the high- or low-pressure turbine side. Both solutions were used in the past, but both increased costs and the turbo set complexity. A gear had to be applied because of the difference in the speed of the turbo set and the pump. As these solutions involved a bigger investment outlay and higher operational costs, they are no longer used. Their substantial drawback was also the forced and extremely unfavorable location of the feed pump in the powerhouse which caused additional construction problems.

Condensing turbines to drive boiler feed pumps are usually fed with the steam from the outlet of the intermediate-pressure turbine or steam from the intermediate-pressure turbine bleeds, and they give steam to the main or additional condenser. The amount of steam needed to drive the condensing turbine reduces the mass flow through the low-pressure turbine and can contribute to a reduction in the total discharge loss and, consequently, reduce the heat consumption per unit. Therefore, an application of a condensing turbine to drive the feed pump is more effective if the low-pressure turbine is heavily loaded.

Under a low load of the unit, the steam mass flow through the low-pressure turbine is usually not big enough which causes the danger of a substantial increase in the temperature of the blades. As a rule, the problem is solved by increasing the steam mass flow from the intermediate- to the low-pressure turbine. Under low loads of the unit, there may then be not enough steam fed into the turbine driving the feed pump which necessitates an introduction of additional steam into it. This may cause mechanical problems because the turbine driving the boiler feed pump was designed for lower parameters of the incoming steam. In consequence, additional facilities are required for the system, which increase the investment costs of the plant. Another option in this situation is a transition to an electric drive of the boiler feed pump.

The large mass flow of the steam needed to satisfy the pump›s demand for the driving power causes the volume of the outlet steam mass flow to increase significantly, which results in a large annulus area and, consequently, in high blades of the last turbine stage. In the case of high power units, the relatively high rotation speeds of the boiler feed pump (up to as much as 7, 300 rotations per minute) are not suitable for the condensing steam turbine. Due to the high rotation speed required by the pump, the peripheral velocities of the last stages differ considerably from the optimum values that ensure the highest efficiency. Combining high peripheral velocities with high blades often causes mechanical problems. To avoid them, some manufacturers offer turbines with the flow separated in the last stage into two stages with a smaller diameter. This sometimes solves mechanical problems with high blades, but due to the flow losses resulting from the separation of the steam mass flow, no improvement in efficiency is achieved. In fact, the efficiency of condensing turbines which drive boiler feed pumps is often by 6% to 7% points lower than that of the main steam turbine [4]. Some manufacturers offer reaction-type condensing turbines with a fully separated flow with the steam inlet in the middle and symmetrical outlets. This leads to the rotor elongation, which is not a problem in the case of reaction constructions because they solve the mechanical problems related to long blades and ensure an efficiency level which is 2% to 3% points, different from the efficiency of the main turbine [4]. However, such machines are very expensive. In recent years, both the power capacity of available electric motors and the efficiency of low-pressure turbines and drive control have increased substantially. Table 1 presents example efficiency values for turbine and electric drives. The efficiency of modern electric motors is much better than the one of motors coming from earlier batches which is still used in electric power stations. The efficiency of the motors offered today exceeds 97%, while older motors reached the efficiency level in the range of 91% to 94%. Moreover, the characteristic curves of the efficiency of modern motors have a flat course, which is of utmost importance for operation under a partial load[5]. Hence, the overall efficiency related to electric motors driving boiler feed pumps has improved. On the other hand, the efficiency of turbines driving BFPs has not changed much. Therefore, in the latest constructions of high capacity power units for supercritical parameters, especially German ones, the system which is most often used is the $2 \times 60\%$ or $2 \times 50\%$ one, with an electric drive and rotation speed controlled by frequency, as this kind of speed control has a fairly flat efficiency characteristic in a wide range of loads compared to the fluid coupling. The adjustment of the pump efficiency is carried out through the control of the frequency and voltage of the current powering the electric motor in order to obtain the required rotation speed. The pump is connected to the driving motor through an elastic coupling to reduce the torsional harmonic forces transmitted by the pump. Changing rotation speed is the most effective and economical way of adjusting the pump efficiency. Adjustments made to rotation speed cause only a slight decrease in the pump efficiency compared to operation with fixed revolutions. A change in rotation speed involves a change in the pump characteristic. This type of control allows a reduction in electricity consumption in a wide range of loads. Usually, frequency-controlled synchronous motors are used for such solutions. The rotation speed of a synchronous motor can be reduced from the nominal value to very low speeds by reducing the frequency and voltage of the electric current powering the motor [5].

Table 1: Comparison of boiler feed pump drive efficiency [6]

Efficiency	Condensing turbine	Electric drive
Low-pressure turbine	-	88-92
Generator	-	98.5
Transformer	-	99.5
Motor	-	97
Drive control	-	95-96
TOTAL	83	74-84

St pczy ska et al.St pczy ska et al. International Journal of Energy and EnvironmentalEngineering 2012 3:3 doi:10.1186/2251-6832-3-3.

The improvement in the reliability and availability of the pumping engine components, which is a part of the general technological progress, has caused that the availability of the 2 × 60% (2 × 50%) systems driven with electric motors is only slightly different from the values achieved by the EM 3 × 50% or BPT 100% + EM 30% configurations. On the other hand, the investment expenditures for the EM 2 × 60% (EM 2 × 50%) systems are much lower compared to the other two[6]. For these reasons, in many recent constructions of high capacity power units, it was the pump driven by a frequency-controlled electric motor that was selected, not the pump driven by the turbine. Despite the electric power losses in the generator, transformer, supply network, frequency converter and motor, this system is sometimes more efficient than the feed pump driven by a condensing turbine.

Alternative Options of the Boiler Feed Pump Drive

The assessment of the possibility of a further improvement in the net efficiency of electricity generation in conventional coal-fired units for ultra-supercritical parameters calls for consideration of options of the boiler feed pump drive other than the most often used ones, namely those powered by the electric motor and by the condensing turbine.

One of them is an application of the aforementioned backpressure turbine or extraction-backpressure turbine fed with outlet steam from the high-pressure turbine from the cold reheat steam line. The outlet steam together with the driving turbine bleeds (if any) would condense in low-pressure regenerative exchangers. Owing to the fact that outlet steam also gives up heat in the regenerative system, the extraction-backpressure turbine does not require a costly system of the condenser and cooling water which usually is the case with the condensing turbine driving the main feed pump.

An interesting option could be the implementation of the 'master cycle' (MC) solution patented by the company Elsam Engineering (Fredericia, Denmark), which consists in shifting all the intermediate-pressure turbine bleeds to an auxiliary extraction-backpressure turbine referred to as 'tuning turbine' which is also fed with the steam from the high-pressure turbine outlet [7-9]. The auxiliary turbine drives the main boiler feed pump and, optionally, an auxiliary generator which could carry away the power surplus generated by the system to the unit transformer through separate, additional primary coils. During the start-up or holdup of the unit, the 'synchro-self-shifting' clutch is open and the auxiliary turbine does not work. The auxiliary generator is switched to motor operation and drives the main feed pump, which renders the start-up feed pump driven by an electric motor unnecessary. The use of an additional generator on a single shaft together with the pump and the turbine would require revolution control to maintain a steady speed on the part of the generator.

A solution like this has many advantages because a substantial rise in the temperature of steam, especially reheated steam, increases the problem of the difference between the bleed steam and the saturation temperatures in regenerative heaters fed with bleeds from the intermediate-pressure turbine, which are situated right after the steam reheater. This causes material and thermodynamic problems because, with the rise in the mean difference in temperature of the heat-exchanging agents, the exergy losses rise too which leads to a reduction in the process effectiveness and to an increase in the heating steam consumption. By keeping pressures in the bleeds of the auxiliary turbine at the same level as in the intermediate-pressure turbine, the temperature difference in the exchangers will be much lower (the outlet of this turbine is even in the area of saturated steam). Another advantage of the master cycle is that the construction of an intermediate-pressure turbine without bleeds is much simpler and cheaper. This is essential in view of the need to use expensive construction materials. Lower temperatures of the steam in the bleeds from the driving turbine, compared to the intermediate-pressure turbine, result in an increase in the mass flow from them. This causes a reduction in the steam mass flow to the reheater, i.e., lower costs of the reheated steam pipelines and of the reheater in the boiler. The steam mass flow from the main turbine into the condenser is also smaller which decreases heat losses of the cycle.

Yet, another solution is the use of a backpressure turbine working in parallel with the intermediate-pressure turbine. The driving turbine is fed with steam from a hot reheat steam line. Outlet steam from the turbine is in turn introduced into the inlet of the low-pressure turbine. Such a configuration limits the problem of too little steam fed to the low-pressure turbine at low loads of the unit. What can be a problem, however, are the high parameters of the inlet steam (identical to those at the inlet to the intermediate-pressure turbine), which is related to the use of expensive construction materials.

An Analysis of Selected Configurations of the Boiler Feed Pump Drive

In order to assess the impact of the boiler feed pump drive configuration on the net efficiency of electric power generation, balance calculations for a unit for ultra-supercritical parameters were carried out. The subject of the study was a conceptual coal-fired unit with a net electric power capacity of 900 MW. The basic parameters of the unit are listed in Table 2. The system has a single inter-stage steam reheat, a four-stage low pressure regeneration and a three-stage high pressure regeneration, plus a steam desuperheater.

Table 2: Basic parameters of the conceptual 900-MW unit

Live steam pressure	30 MPa
Live steam temperature	650°C
Reheated steam pressure	6 MPa
Reheated steam temperature	670°C
Pressure in the condenser	0.005 MPa

Feed water temperature	310°C
Net electric power	approx. 900 MW

Stępczyńska et al.Stępczyńska et al. International Journal of Energy and EnvironmentalEngineering 2012 3:3 doi:10.1186/2251-6832-3-3.

Five configurations of the boiler feed pump drive were analyzed:

- A frequency-controlled EM (Figure 1)
- A CT fed with steam extraction from the immediate-pressure turbine (Figure 2)
- An EBPT fed with steam from a cold reheat steam line with one bleed and an outlet directed to low pressure regenerative exchangers (Figure 3)
- A BPT fed with steam from a hot reheat steam line operating in parallel with the intermediate-pressure turbine (Figure 4)
- An application of the idea of the MC: an extraction-backpressure turbine fed with outlet steam from the high-pressure turbine with bleeds and an outlet directed to regenerative exchangers which, in a classic system, are fed from the intermediate-pressure turbine (Figure 5).

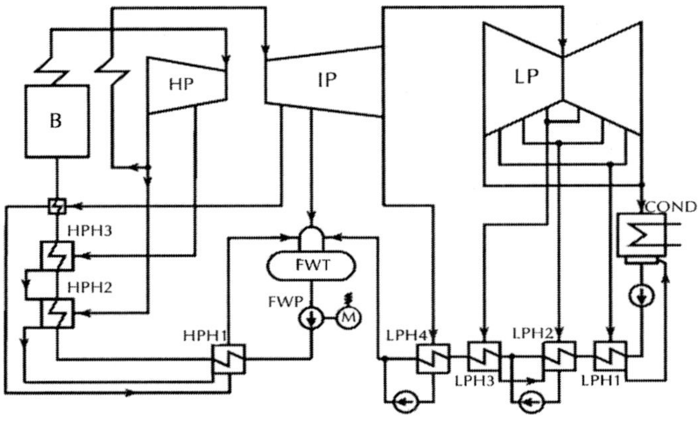

Figure 1: Diagram of the thermal cycle of a 900-MW unit with feed pump driven by EM. B, boiler; HP, high-pressure turbine; IP, immediate-pres-

sure turbine; LP, low-pressure turbine; HPH3, HPH2 and HPH1, high-pressure feed water heaters; FWP, feed water pump; FWT, feed water tank + deaerator; LPH4, LPH3, LPH2 and LPH1, low-pressure feed water heaters; COND, condenser.

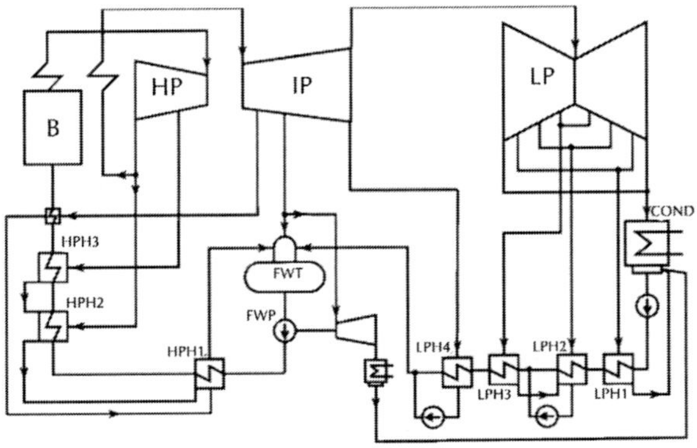

Figure 2: Diagram of the thermal cycle of a 900-MW unit with feed pump driven by CT. B, boiler; HP, high-pressure turbine; IP, immediate-pressure turbine; LP, low-pressure turbine; HPH3, HPH2 and HPH1, high-pressure feed water heaters; FWP, feed water pump; FWT, feed water tank + deaerator; LPH4, LPH3, LPH2 and LPH1, low-pressure feed water heaters; COND, condenser.

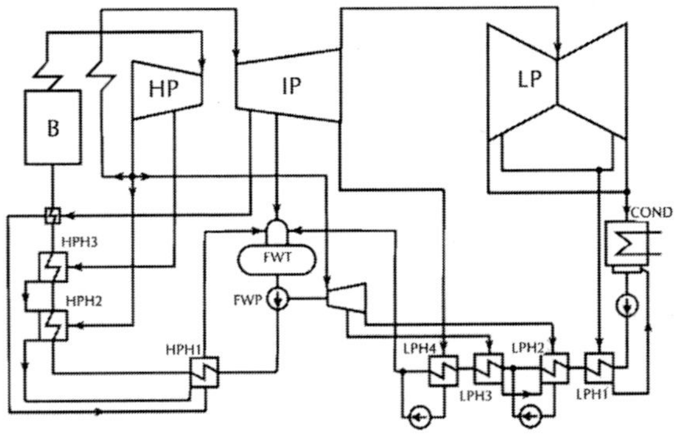

Figure 3: Diagram of the thermal cycle of a 900-MW unit with feed pump driven by EBPT. B, boiler; HP, high-pressure turbine; IP, immediate-pres-

sure turbine; LP, low-pressure turbine; HPH3, HPH2 and HPH1, high-pressure feed water heaters; FWP, feed water pump; FWT, feed water tank + deaerator; LPH4, LPH3, LPH2 and LPH1, low-pressure feed water heaters; COND, condenser.

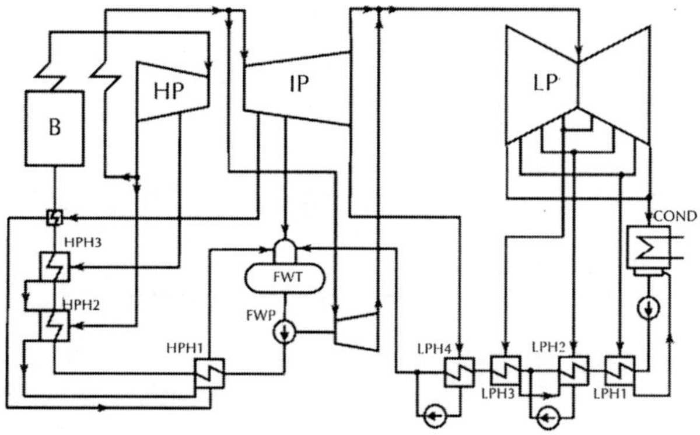

Figure 4: Diagram of the thermal cycle of a 900-MW unit with feed pump driven by BPT. B, boiler; HP, high-pressure turbine; IP, immediate-pressure turbine; LP, low-pressure turbine; HPH3, HPH2 and HPH1, high-pressure feed water heaters; FWP, feed water pump; FWT, feed water tank + deaerator; LPH4, LPH3, LPH2 and LPH1, low-pressure feed water heaters; COND, condenser.

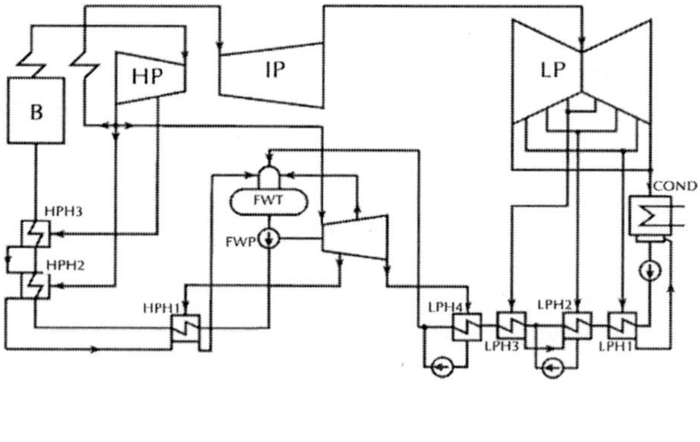

Figure 5: Diagram of the thermal cycle of a 900-MW unit. With feed pump driven by extraction-backpressure turbine: the idea of the MC.

B, boiler; HP, high-pressure turbine; IP, immediate-pressure turbine; LP, low-pressure turbine; HPH3, HPH2 and HPH1, high-pressure feed water heaters; FWP, feed water pump; FWT, feed water tank + deaerator; LPH4, LPH3, LPH2 and LPH1, low-pressure feed water heaters; COND, condenser.

Figures 1, 2, 3, 4 and 5 present simplified diagrams of the thermal cycle of a 900-MW unit for the boiler feed pump drive configurations under consideration.

Methods and Assumptions in the Cycle Analysis

The calculations of the cycle for selected configurations of the boiler feed pump drive ware conducted with the software Gate Cycle v. 5.40 (GE Company, Billerica, MA, USA). The software is used for balance calculations of electric power systems and makes it possible to model their operation under various loads.

The analysis of the cycle was conducted with an assumption that for each unit configuration the nominal mass flow of live steam was identical. The efficiencies of the groups of turbine stages as well as the pressures in corresponding bleeds in individual unit configurations have the same values so do all the live steam parameters, the pressure in the condenser, the temperature of the water feeding the boiler and the boiler efficiency.

The calculations were made for the nominal mass flow of live steam produced by the boiler and for the partial mass flow of 75% and 50%. For the needs of the analysis, it was also assumed that, for each configuration, the system had the same boiler feed pump that ensured a 100% efficiency of the steam boiler, and the booster pump had a separate electric drive. In the case of the variants with the turbine drive of the feed pump, the turbine generates power needed only to drive the pump and does not generate extra electricity.

In order to calculate the net electric power and the net efficiency of the unit, an assessment was made of the demand for electric power of all the basic auxiliary equipment: the boiler feed pumps: both booster and main (for the EM variant), the condensate pumps, the pump of the water cooling the condenser, the air and flue gas fans, and the coal pulverizers. In each variant, identical operation and efficiency characteristics were adopted for corresponding facilities.

RESULTS AND DISCUSSION

The paper presents a comparison of the variants of the boiler feed pump drive for a conceptual 900-MW coal-fired unit for ultra-supercritical parameters. Table 3 shows the values of electric power needed to drive the feed pump for individual options depending on the live steam mass flow. In the electric motor option, the demand for electric power accounts for energy losses in the transformer, the supply network, the electric motor and in the frequency converter. For the remaining options with the feed pump turbine drive, the demand for electric power to drive the pump was calculated as the difference between the unit gross electric power for the option with an electric motor and the unit gross electric power for the other variants. The variant with the BPT shows the lowest demand for power for the 100% and 75% mass flow of live steam; for a smaller load of 50%, the CT requires the least power. However, for the EM, CT and BPT variants, the values are similar. A considerable reduction in the generated gross electric power occurred for the EBPT and the MC variants. This is because, in both of these cases, a relatively large steam mass flow is directed to the turbine driving the feed pump (for the MC option, it is more than 15% of the live steam mass flow), and it does not work in the main turbine. As the driving steam is taken into the turbine from a cold reheat line, it is not reheated in the boiler. Consequently, the amount of heat delivered into the cycle is reduced in comparison with the other variants.

Table 3: Demand for electric power to drive the feed pump

Drive variant of the feed water pump	**Live steam mass flow/ nominal live steam mass flow**		
	50%	75%	100%
Electric motor (EM)	9,844 kW_e	18,364 kW_e	32,660 kW_e
Condensing turbine (CT)	9,261 kW_e	18,239 kW_e	32,947 kW_e
Extraction-backpressure turbine (EBPT)	16,322 kW_e	28,723 kW_e	47,703 kW_e
Backpressure turbine (BPT)	10,041 kW_e	17,821 kW_e	30,169 kW_e
The idea of the Master Cycle (MC)	26,674 kW_e	44,441 kW_e	66,756 kW_e

Stępczyńska et al.Stępczyńska et al. International Journal of Energy and EnvironmentalEngineering 2012 3:3 doi:10.1186/2251-6832-3-3.

Table 4 presents the unit consumption of the fuel chemical energy per unit of the generated net energy and the unit net electric power for three values of the live steam mass flow: 100%, 75% and 50% of the nominal mass flow. For the nominal load of the unit, the lowest unit consumption of the fuel chemical energy, i.e., the highest net efficiency of electric power generation was achieved for the MC variant, whereas the least efficient was the EM option. Figure 6 presents in the graph form the changes in the values of the unit net efficiency depending on the variant of the boiler feed pump drive. The lines in the graph correspond to the set values of the live steam mass flow produced by the boiler. For lower values of the live steam mass flow, the CT variant turned out to be much better. For the MC and EBPT configurations, there is a bigger fall in the unit efficiency as the load decreases than for the others. This is probably due to the fact that turbines in this configuration have a large impact on the operation of regenerative exchangers where the steam from their bleeds and outlets is directed. Whenever the mass flow and the pressure of the generated steam decrease, the demand for power of the boiler feed pump falls substantially. Owing to that, the mass flow of the steam reaching the driving turbine also decreases significantly, and as a result, the pressures and steam mass flow at its bleeds and outlet decrease as well. Less steam flows into regenerative exchangers, so less heat is given up to the water that is being heated. The situation would be different if the steam mass flow feeding the turbine was increased, and the resulting surplus of power was given up in the auxiliary generator. However, as it was mentioned before, there is a problem with the difference in rotation speed between the feed pump and the generator. The pump operates at high speeds which change according to load, whereas the generator has to maintain a steady synchronous rotation speed.

Table 4: Unit net electric power (NelN) and unit fuel consumption per unit of generated net energy (qenchN)

Drive variant of the feed water pump	**Live steam mass flow/ nominal live steam mass flow**					
	50%		**75%**		**100%**	
	N_{elN} (kW_e)	**q_{enchN}(kJ/kWh)**	**N_{elN}(kW_e)**	**q_{enchN}(kJ/kWh)**	**N_{elN} (kW_e)**	**q_{enchN} (kJ/kWh)**
Electric motor	478,980	7,847	701,798	7,684	908,255	7,636
Condensing turbine	479,719	7,835	702,200	7,679	908,461	7,632
Extraction-back-pressure turbine	472,875	7,849	692,001	7,677	894,126	7,620
Back-pressure turbine	479,001	7,819	702,686	7,656	911,273	7,608
The idea of the Master Cycle	462,999	7,854	676,888	7,666	875,774	7,581

Stępczyńska et al.Stępczyńska et al. International Journal of Energy and EnvironmentalEngineering 2012 3:3 doi:10.1186/2251-6832-3-3.

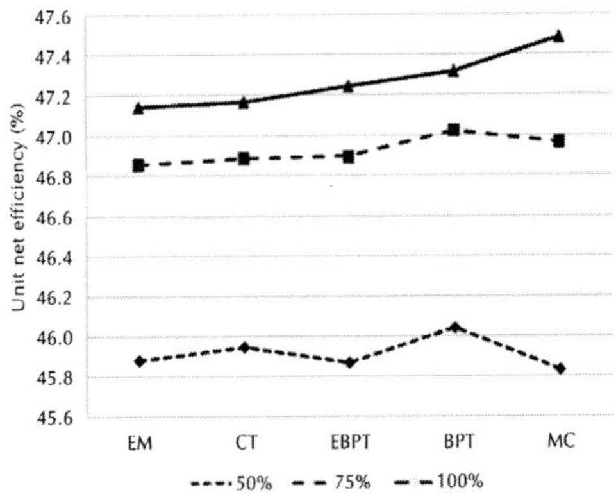

Figure 6: Net efficiency of a 900-MW unit electric power generation.For individual feed pump drive options depending on the live steam mass flow. EM, electric motors; CT, condensing turbine; EBPT, extraction-back-pressure turbine; MC, master cycle.

CONCLUSIONS

The analysis showed that for the assumptions adopted for the calculations of a 900-MW ultra-supercritical unit, the highest net efficiency of electric power generation with a nominal steam mass flow had been achieved for the MC variant. It is by 0.34% point higher than the one obtained for the feed pump driven by an electric motor. According to Blum et al. [7], the application of such solutions gives much better results if they are combined with the use of dual steam reheat, where the problem of temperature accumulation in regenerative exchangers is more significant.

For smaller mass flows of generated steam, the cycle with a BPT features the highest efficiency. This type of solution might, however, entail higher investment expenditures due to very high parameters of the inlet steam (reheated steam). The differences in the unit net efficiency for variants using typical drive configurations, EM or CT, are very slight. The load schedule of the unit should also have a considerable influence on the selection of the boiler feed pump drive. Assuming a longer operation of the unit under lower loads would exclude the application of the EBPT and MC variants due to a substantial decrease in the unit efficiency with load.

The choice of the optimum option of the boiler feed pump drive has a significant meaning from the point of view of reliability and availability of a power unit, the amount of investment expenditures and the net efficiency of electric power generation. Due to that, for each new investment, a detailed technical-economic analysis of the possible solutions has to be conducted.

AUTHORS' CONTRIBUTIONS

KS carried out the calculations for all 900-MW power unit configurations, participated in the sequence alignment and drafted the manuscript. SD conceived of the study and participated in its design and coordination, and helped draft the manuscript. HŁ participated in the design of the study and helped draft the manuscript. All authors read and approved the final manuscript.

ACKNOWLEDGMENTS

The results presented in this paper were obtained from research work co-financed by the National Centre for Research and Development within the framework of contract SP/E/1/67484/10—Strategic Research Program—Advanced Technologies for Energy Generation: Development of a technology for highly efficient zero-emission coal-fired power units integrated with CO_2 capture.

REFERENCES

1. Buchta, J, Pawlik, M, Oziemski, A, Kotlicki, T, Wawszczak, A: Potrzeby własne bloku nadkrytycznego. In: Kosman G, Rusin A, Taler J, Pawlik M (eds.) Zagadnienia projektowania i eksploatacji kotłów i turbin do nadkrytycznych bloków w glowych, pp. 390–394. Politechniki l skiej, Gliwice (2010)

2. Laudyn, D, Pawlik, M, Strzelczyk, F: Elektrownie, Naukowo-Techniczne, Warszawa (2005)
3. Westhuizen, W, Cattaert, T: Power station pump selection: part 2. Worlds Pump.. 1, 14–17 (2010)
4. Enery-Tech.Consider dual-driven boiler feed pumps for coal plants with capacity greater than 500 MW:
5. Szulc, Z: Regulowany nap d elektryczny pompy wody zasilaj cej o podwy szonej pewno ci zasilania. (2008) Elektro. info. http://www.elektro.info.pl/artykul/id2534,regulowany-naped-elektryczny-pompy-wody-zasilajacej-o-podwyzszonej-pewnosci-zasilania. Accessed 10 January 2011
6. www.voilth.pl. Zastosowanie przekładni hydrodynamicznych Voith w nap dach pomp wody zasilaj cej nowych bloków energetycznych. http://www.voith.pl/index.php?id_strony=302&id=241&idmain=26. Accessed 10 January 2011
7. Blum, R, Bugge, J, Kjaer, S: USC 700°C power technology—a European success story. VGB PowerTech.. 26, 4 (2009)
8. Blum, R, Bugge, J, Kjaer, S: Development of a PF fired high efficiency power plant (AD700). Proceedings of the Riso International Energy Conference "Energy Solutions for Sustainable Development, Riso R-1608 (EN), Denmark (2007)
9. Pawlik, M: Elektrownie opalane w glem w wietle polityki ograniczania emisji CO2. Biuletyn Techniczno-Informacyjny Zarz du Oddziału Łódzkiego Stowarzyszenie Elektryków Polskich.2(45), 1–36 (2009) http://sep.p.lodz.pl/biuletyn/sep_2_2009.pdf

Chapter 2

STAR CCM+ CFD Simulations of Enhanced Heat Transfer in High-Power Density Electronics Using Forced Air Heat Exchanger and Pumped Fluid Loop Cold Plate Fabricated from High Thermal Conductivity Materials

Kevin R. Anderson[1], Matthew Devost[1], Watit Pakdee[2], and Niveditha Krishnamoorthy[3]

[1]Nonlinear FEA/CFD Multi-Physics Simulation Laboratory, Department of Mechanical Engineering, California State Polytechnic University, Pomona, USA

[2]Department of Mechanical Engineering, Faculty of Engineering, Thammasat University (Rangsit Campus), Bangkok, Thailand

[3]CD-Adpaco, One Technology Drive, Irvine, CA, USA

ABSTRACT

As telecommunication and RF power electronics applications continue to push the envelope of waste heat dissipation, more and more, we see a need for active thermal control employing forced air electronic cooling fans in unison with pumped fluid loops in order to meet temperature and performance requirements. This research paper presents results of applying Computational Fluid Dynamics (CFD) commercial industry STAR-CCM+ software for heat transfer and fluid flow simulation of a novel heat exchanger/cold plate fabricated from k-core high thermal conductivity material in order to realize thermal control system hardware design for very much applications to very large power density (~1 kW/m^2) electronics packaging scenarios. Trade studies involving different heat exchanger/cold plate materials, as well as various fault scenarios within a mock-up of a typical electronics system, are used to illustrate the upper bounds placed on the convective heat transfer coefficient. Agreement between our present findings and previous research in the field of electronics cooling is presented herein.

INTRODUCTION

The novelty of the current research focuses on a system's level approach to using CFD to examine "cause-and-effect" scenarios in an electronics cooling application. In particular, our thermal control solution uses a very high thermal conductivity material known as "k-core" [1] in order to implement a heat exchanger/cold plate used to cool high power density electronics packaging. Effective heat spreading has been implemented by previous investigators interested in heat transfer augmentation including the work of [2], which used anisotropic heat plate with preferential thermal conductivity to augment thermal performance. In the study of [3], the critical heat flux of a heat spreader was quantified. While in [4], the use of diamond coated materials was researched. This current paper utilizes Computational Fluid Dynamics (CFD) in order to study various causes and effects at a system's level. The CFD work related to electronics currently available in the literature can be broken into the following categories based on level of characteristic thermal/fluid length scale, or overall size of the hardware being analyzed: MEMS level; chip/component level; Printed Circuit Board (PCB) board level; hand-held devices (cell phones, instrumentation, etc.); desk-top computer level; intermediate compact packaging level; rack-mounted server level; very large length scale (i.e. power grid distribution applications, etc.). Herein we are interested in the intermediate compact packaging level.

At the chip/component/PCB level, various researchers have used CFD to study electronics cooling with the use of a heat sink including the work of [5], which implements CFD to characterize the performance of carbon composite and copper heat sinks on the heat transfer in a computer CPU. The work of [6] utilizes CFD to perform trade studies on the modeling of finned heat sinks, while [7] has married experimental and CFD analysis to quantify the effects of heat sinks and base plates on heat transfer behavior. Finally the work of [8] performs CFD and presents heat transfer correlations for the optimization of fin heat sinks. At a desk-top computer level, CFD has been used to study the heat transfer in computer chassis and computer box such as the work in [9] which focuses on computer chassis heat transfer performance. The study of [10] implements CFD to investigate PCB heat transfer in a computer. The study of [11] investigates computer heat transfer and cooling using a heat-pipe. More generic analyses of using CFD to electronic systems are given in the work of [12-14], wherein system's level analysis is performed using CFD to gain insight on heat transfer and thermal performance of various electronic devices, ranging from hand-held devices to desktop computers. The tutorial offered by [15] allows the novice to gain insight on how to use electronics packaging design lay-out strategies to realize more effective heat transfer flow field patterns. In the arena of rack mounted servers, several pertinent CFD based investigations have been carried out over the last decade or so. The most noteworthy of these is the study of [16], wherein a CFD tool is implemented to study the temperature distribution in rack mounted servers. The investigation of [17] uses a liquid cooling thermal management system to enhance the heat transfer in a rack cooling data center application, while the work of [18] outlines a road-map for future CFD work related to rack mounted cooling systems. For the intermediate length scale, the works of [19, 20] address this type of hardware, but the studies only address moderate power density heat transfer thermal control strategies. The work of [19] includes the fidelity of the thermal vias, solder, PCB, and shelf interface. Other applications of using numerical heat transfer and CFD in electronic and avionic systems are given in [21, 22]. Presently, there appears to be a void of CFD simulation databases addressing the so-called intermediate application level. The goal of this current study is to address this absence of literature by simulation of an intermediate sized electronics enclosure, which is subject to very large power densities. This is in an effort to address the ever changing requirements of affording more power in a smaller packaging level.

To date, CFD of intermediate level packaging highpower density enclosures has not been carried out comprehensively. Furthermore, there have been few or no CFD simulations of electronics cooling using CD-Adapco's STAR-CCM+ CFD software in the literature, mainly ICEPAK and FLOTHERM, which have the tools of choice as evidenced in the works of [5, 6, 9, 12-15]. To address this absence in the literature, the current CFD simulations focus on using STAR-CCM+ CFD to simulate heat transfer and fluid dynamic behavior in intermediate sized Telecommunication/Observatory Science/Electronics/RF/Avionics systems fabricated primarily from COTS (Commercially Available Off The Shelf) components (i.e. fans, heat-sinks, cold plates, etc.) to implement the thermal control solution. This current study focuses on cost-effective thermal control system alternative, in contrast to more cost-prohibitive solutions such as those involving heat pipes, nano-fluids and/or phase change materials. The current study uses CFD to quantify the effects of heat transfer augmentation using high thermal conductivity materials in high power density electronics cooling scenarios.

HEAT SPREADER THERMAL TECHNOLOGY

Herein, "k-core" thermal conduction plates available from Thermacore, Inc. [1] are used in conjunction with a pumped fluid loop to illustrate the effects of heat transfer spreading. K-core heat spreader cold plates afford highefficiency heat transfer in the absence of moving parts, and are invaluable in applications where space, volume, seamless hardware integration and weight constraints dictate thermal design solutions. Physically, k-core base heat spreader cold plates transfer heat from sources such as high power dissipating electronic components, rejecting it to ambient air (through forced or natural convection) or liquid coolants (water, glycol/water mixtures, dielectric fluids, polyalphaolefin or PAO). These types of k-core cold plates are well suited for operating in harsh environments, including extreme temperatures/humidity, shock and vibration. For cooling high-power density applications, such as semiconductors, lasers, power generation, medical equipment, transportation, military electronics, etc., and other demanding applications where air cooling is insufficient, liquid cooling (pumped fluid loop) is essential. In this case Thermacore k-core based liquid cooled cold plates offer a feasible thermal control system solution. Thermacore's custom liquid cooling component cold plates include tube-in-plate, aluminum vacumnbrazed and copper brazed types for various applications. Tube-in-plate cold plate materials consist of copper or stainless steel tubes pressed into a channeled aluminum or copper extrusion or machined plate. Thermacore also offers copper cold plates that incorporate vertical fin (micro-channel) technology or, for higher performance, our specialized powdered metal construction. Thermacore's custom liquid-cooled assemblies are designed specifically for each application's unique thermal/mechanical requirements. The physical concept of Thermacore's k-core heat transfer technology is illustrated in Figure 1.

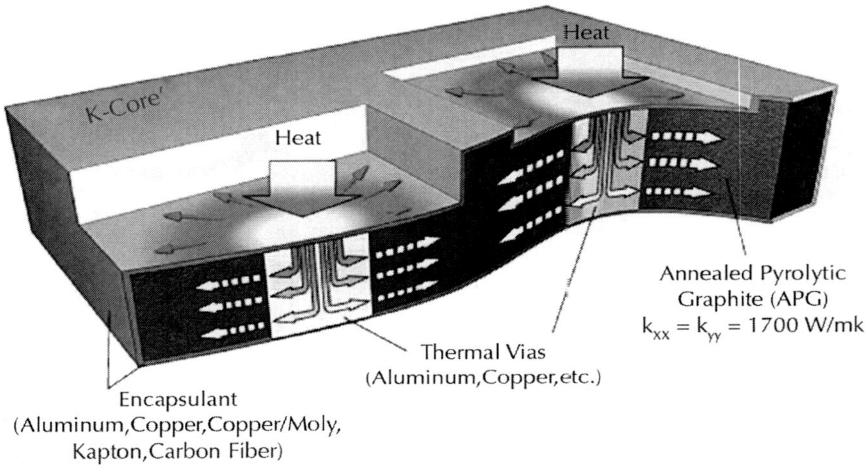

Figure 1: Thermacore k-core technology [1].

Figure 1 shows the various thermal technologies offered by Thermacore, Inc. Using Encapsulated Annealed Pryrolytic Graphite (APG) allows one to manufacture a device such as a heat spreader with heat sink integrated on one face while a labyrinth of tubing for a pumped fluid loop can be fabricated on the opposing face.

The encapsulant of Figure 1 sets the Coefficient of Thermal Expansion (CTE) and other structural properties. The encapsulant material is selected to satisfy requirements such as ease of integration into the overall system. Since the APG is anisotropic in thermal conductivity, i.e. the x-component, and y-component of thermal conductivity is very large with respect to the y-component of the thermal conductivity, the APG acts as an excellent mechanism with which to realize heat transfer enhancement via preferential heat spreading. Figure 2 illustrates the temperature dependency of the thermal conductivity of APG. From Figure 2 it can be seen the APG has its largest thermal conductivity at 150 K (−123 C) where k= 2800 W/m-K making is a valuable thermal heat transfer material for spacecraft applications. For electronics cooling applications where the nominal temperature is on the range of 350 to 400 K (77 C to 127 C) the APG has a thermal conductivity of k = 1500 W/m-K (as compared to Copper have k = 400 W/m-K, and Aluminum = 200 W/m-K). In this current study we investigate the use of APG k-core technology for temperatures in the range of 266 K to 373 C (29 C to 100 C), k = 1600 W/m-K to k = 1100 W/m-K.

SYSTEM HARDWARE DESCRIPTION

Figure 3 shows the layout of the electronics being investigated in this paper. The configuration of Figure 4 was selected in order to mimic an intermediately sized high power density power electronics application. The overall system consists of a very large housing with eight identical sub-modules arranged in order to minimize overall system volume. The sub-modules were each populated with various components, totaling twelve each per submodule. These components were left to be very generic in order to facilitate our trade study involving simulation of system level electronics using CFD.

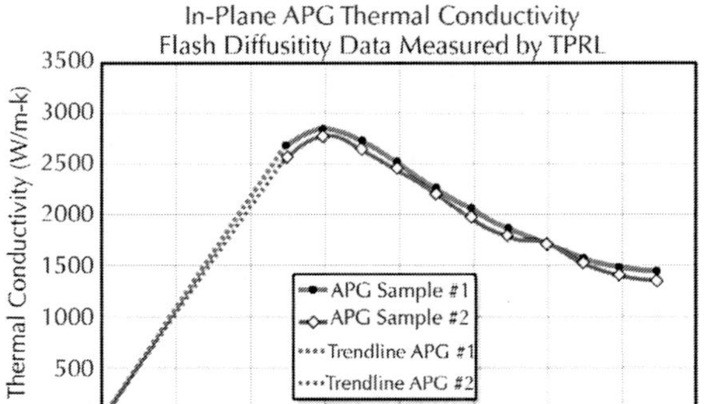

Figure 2: Thermal conductivity for APG material [1].

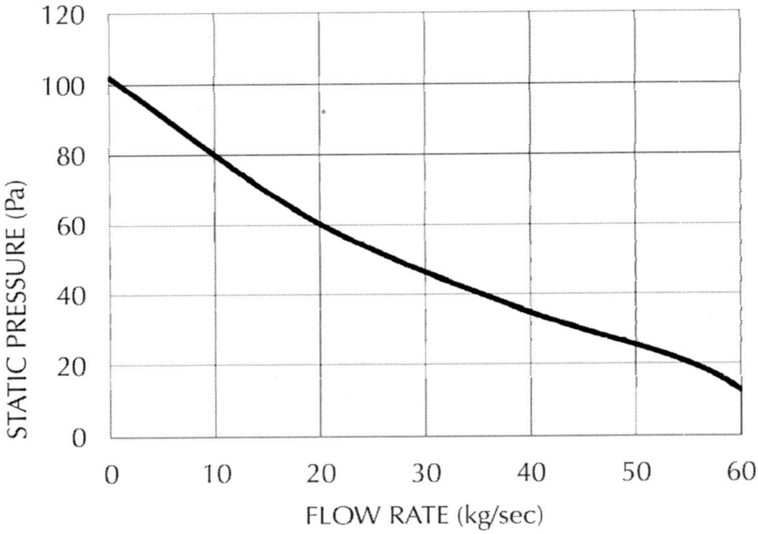

Figure 3: Fan curve specifications for thermal control hardware simulations.

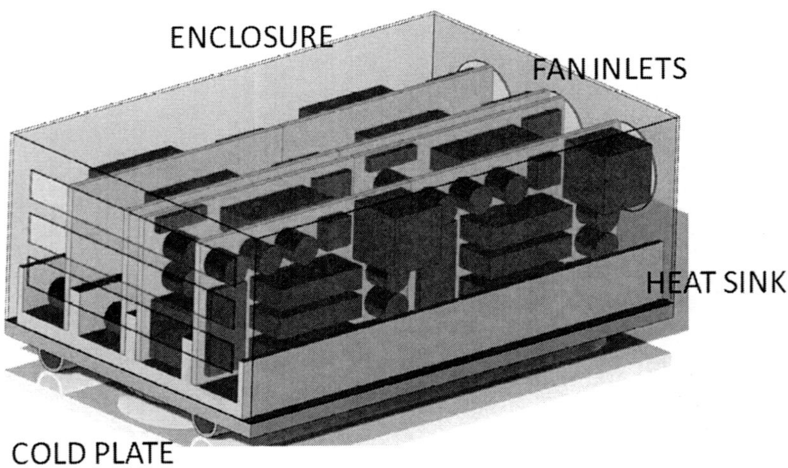

Figure 4: Electronics System Components.

As seen in Figure 4, the thermal control system is comprised of three major components; the heat sink, the cold plate and the forced convection fans/housing enclosure. The majority of heat transfer removal capacity is accomplished via the forced convection air flow of the fans, while the cold plate serves to hold a constant boundary temperature in order to act as a heat sink. This enclosure measures overall dimensions W*L*H = 0.5 m by 1.0 m by 0.5 m (19.7 inches by 39.4 inches by 19.7 inches, thus it is roughly the size of a foot-locker, smaller than a typical rack-mounted server, and larger than a typical PC workstation tower. As is traditionally done in practice in electronics manufacturing, no particular attention was paid to the exact layout of the overall system, i.e. the eight sub-modules were placed into the overall enclosure in a staggered configuration. This of course is not optimum, since some components on some of the submodules will be exposed to fans while others will not. The effects of material selection for the manufacturing of the heat exchanger and fluid loop cold plate has profound ramifications on the optimization of the heat transfer enhancement of the electronics package as a whole. Thus, the current study focuses on the effects of using the previously motivated APG k-core high thermal conductivity material in the fabrication of an integrated heat exchanger, cold plate on the convection heat transfer coefficient (aka h-value) in electronics cooling applications.

The CFD simulations presented herein utilized fully 3-d conjugate heat transfer; forced external air convection; forced internal cooling loop; solid conduction; surface to surface gray body radiation, turbulent. To mimic typical COTS cooling fans, the fan-curve of Figure 3 below was used in the simulations. Figure 3 gives fancurve characteristics for a fan rated at 30 kg/sec with 50 Pa static pressure (60 cfm with 0.2 inches W.G.).

The parametric trade space investigated over the course of our investigation is shown below in Table 1.

The electronics power listed as 500 to 1000 W in Table 1 was accomplished as per the following example, for a 720 W system, each component was assign 7.5 W, thus 7.5 W/component*12 components/sub-module*8 sub-modules/housing*1 housing = 720

W. Using the footprint of 1 m*0.5 m of the cold plate, this affords and electronics power density of 1440 W/m^2. According to Figure 5, the viable technologies for a tolerated temperature delta of 20 C involves direct air, forced convection, while for temperature deltas on the order of 100 C, a hybrid of forced air and natural convection in addition to radiation is recommended. However, in spacecraft and terrestrial applications of high power density heat transfer there is no gravity, thus natural convection cannot be relied on. Additionally, radiation effects in the presence of forced convection have been found to be minimal by prior CFD investigations [9, 12, 14].

Table 1: Parametric trade space for k-core thermal control system study

Parameter	Range
Electronics Power (W)	500 - 1000
Electronics Power Density (Kw/m2)	1 - 2
Inlet Air Flow (m/s)	5 - 15
Inlet Air Temperature (°C)	10 - 20
Water Inlet Temperature (°C)	10 - 20
Water Inlet Flow Rate (kg/s)	0.10 - 0.20
Heat Sink Thermal Conductivity (W/m-k)	200 - 1100
Cold Plate Thermal Conductivity (W/m-k)	200 - 1100

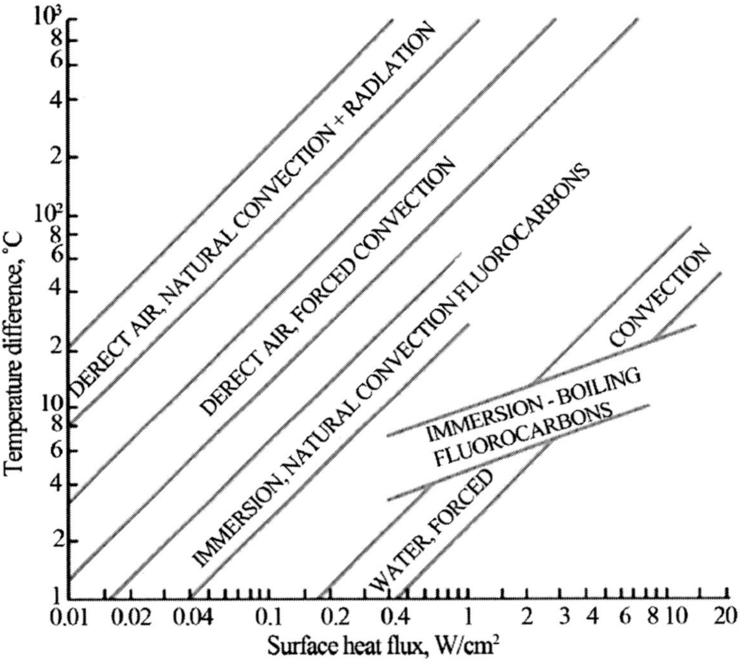

Figure 5: Roadmap of thermal technologies [23].

Thus a hybrid solution of forced convection and highly thermally conductive heat sink/pumped fluid looped thermal control system architecture is proposed herein to address spacecraft and terrestrial and avionics applications which continue to push the envelope of power size to area ratio to the extreme limits of today's capable "stateof-the-art" technologies.

CFD MODEL DESCRIPTION

The CFD package STAR-CCM+ from CD-Adapco was used to import the system geometry of Figure 3. The workflow in STAR-CCM CFD modeling is mush the same as in other packages, such as ANSYS-FLUEN, Flotherm, NX-Space Systems Thermal, COMSOL, etc. First the geometry is imported, then materials are assigned to components, boundary conditions are assigned, mesh is generated, solvers/models are selected, the equations of motion are then solved and results are post-processed. For the simulations outlined herein the STAR-CCM+ CFD software uses Polyhedral Unstructured Mesh with Prismatic Boundary Layer enhancement. The mesh used in the CFD simulations presented herein is shown below in Figure 6.

The software package STAR-CCM+ allows the user to specify the regions (parts) as well as the thermal interface between each part. The cooling fans are modeled in STAR-CCM using an in-place interface whereby in Figure 5 this is illustrated with the dark blue circles situated at the northwest portion of the diagram. These in-place interface fans allow the used to assign a fan-curve of the form of that given in Figure 5 to be keyed in as a polynomial curve fit. STAR-CCM+ also allows the user to model the swirl of the fan, however for this system's level study, this option was found to be too computationally expensive, since we have three fans at the inlet of the housing to simulate. The impact of this modeling assumption if felt to be minimal, since our primary objecttive herein was to simulate the flow of air through the entire enclosure with a fan rated at a certain volumetric flow rate and static pressure drop, the use of interface fans is warranted. The study of swirl around the vicinity of the fans is not within the current scope of investigation.

Grid Independence Study

Typical cell count for the system level CFD model was on the order of 2.5 million cells. Five prismatic boundary layer cells were employed to resolve the boundary layer at all solid-fluid interfaces. Various trades were run in order to determine the optimum mesh size, for 2.5 million polyhedral cells, a base element size of 0.01 m was used, mesh independence studies centered around this value were performed, i.e. the base element size was varied from 0.005 m (more cells, highest fidelity mesh) to 0.02 m (fewer cells, coarsest mesh). A monitor point was selected in the model, and the primitive variables were monitored as the cell size was halved, resulting in the choice of 0.01 m (2.5 million cells) as the baseline model size. CPU run-time on a 64-bit quad-core DELL PC workstation was on the order of six hours.

Governing Equations

Headings, The governing equations solved in STARCCM+ via the Finite Volume Method for our laminar, conjugate heat transfer simulations are the incompressible Navier-Stokes equations stated below in continuous integral form

$$\frac{\partial}{\partial t}\int_V \boldsymbol{W}\chi \mathrm{d}V + \oint [\boldsymbol{F}-\boldsymbol{G}]\cdot \mathrm{d}\boldsymbol{a} = \int_V \boldsymbol{H}\mathrm{d}V \tag{1}$$

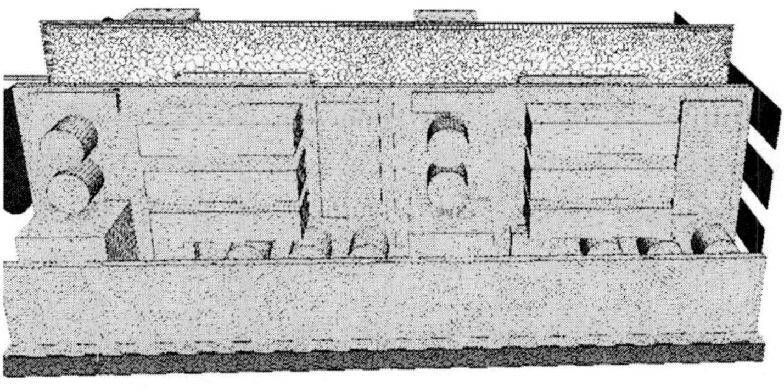

Figure 6: CFD mesh used in simulations.

$$\boldsymbol{W}=\begin{Bmatrix}\rho\\ \rho\boldsymbol{u}\\ \rho E\end{Bmatrix}, \boldsymbol{F}=\begin{Bmatrix}\rho(\boldsymbol{u}-\boldsymbol{u}_g)\\ \rho(\boldsymbol{u}-\boldsymbol{u}_g)\otimes\boldsymbol{u}+p\tilde{I}\\ \rho(\boldsymbol{u}-\boldsymbol{u}_g)H+p\boldsymbol{u}_g\end{Bmatrix},$$

$$\boldsymbol{G}=\begin{Bmatrix}0\\ \tilde{T}\\ \tilde{T}\cdot\boldsymbol{u}+\dot{\boldsymbol{q}}''\end{Bmatrix}, \boldsymbol{H}=\begin{Bmatrix}S_u\\ \boldsymbol{f}\\ S_u\end{Bmatrix} \tag{2}$$

where r, u, E, and p denote the density, velocity, total energy per unit mass, and pressure of the fluid, respecttively. The parameter $\tilde{T}$ is the viscous stress tensor, while $\dot{q}''$ denotes the heat flux vector and u_g is the grid velocity vector. The total energy is related to the total enthalpy via the caloric equation of state

$$E = H - \frac{p}{\rho} \tag{3}$$

where the total enthalpy is the sum of the internal energy and the kinetic energy via

$$H = C_p T + \frac{u^2}{2} \tag{4}$$

For the remaining terms in the Navier-Stokes equation, H denotes the body force vector, W is the vector of conserved quantities F is the vector of inviscid terms, G is the vector of viscous terms, $\tilde{I}$ is the identity tensor, c denotes the porosity, and m is the dynamic viscosity. The constitutive equation modeling heat conduction in the solid bodies is taken to by Fourier's Law stated below in differential form

$$\boldsymbol{q}'' = -k\nabla T \tag{5}$$

The above equations are implemented on an unstructured polyhedral mesh and the resulting algrebraic equations are solved using the algebraic multi-grid method.

RESULTS

The results for temperature distribution, flow vectors and streamlines in our simulations are illustrated in Figures 7-9. The parameters for the simulation were as follows: each component has 7.5 W giving 720 W for 12 components and 8 sub-modules. Additionally, each PCB card had an additional 10 W of power applied to it, thus 80 W for the total package of eight cards. Hence, this simulation is 800 W over a 1 m × 0.5 m footprint giving 1600 W/m^2 = 1.6 kW/m^2, which is in the category of a very high power density application. Figure 7 shows the inlet air coming from the fans at 303 K (30°C) and a heat sink/cooling plate temperature of 283 K (10°C). Figure 8 shows the velocity vectors for the situation of Figure 7, while Figure 9 illustrates streamlines of the flow field for the system of Figure 7.

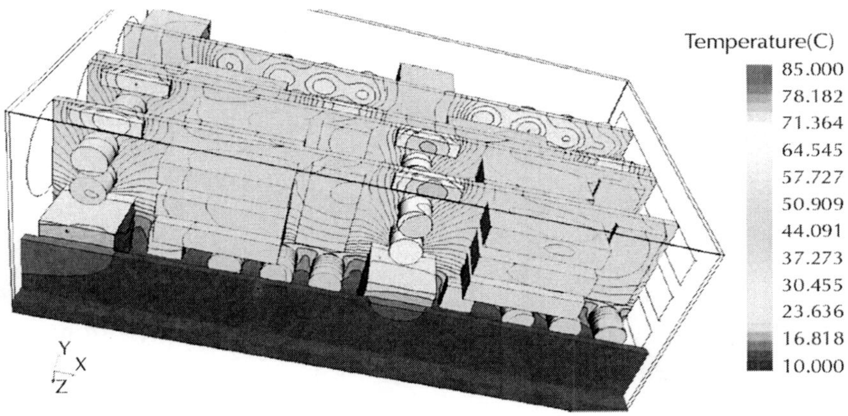

Figure 7: Component temperatures.

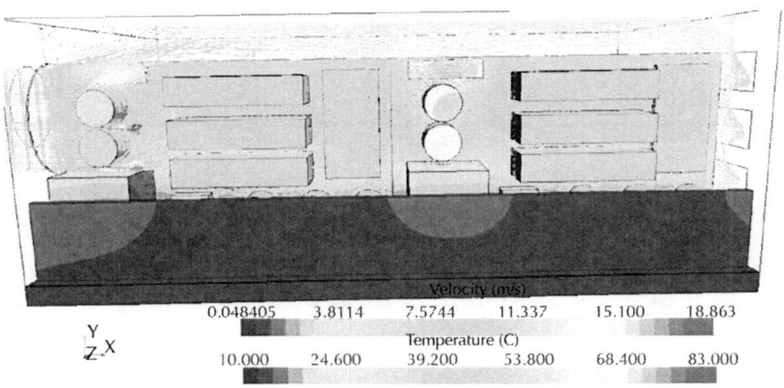

Figure 8: Velocity vectors colored by air speed and electronics components colored by temperature.

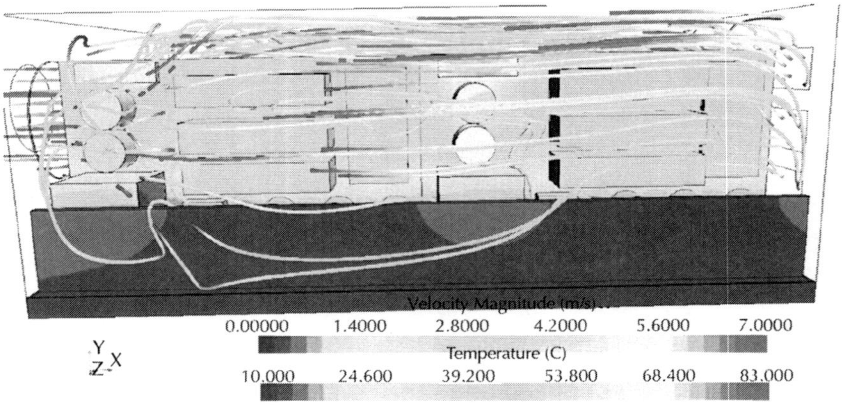

Figure 9: Streamlines colored by air stream speed and electronics components colored by temperature.

The heat sink/cooling plate assemblage serve to keep the baseplate of the system at 10˚C, serving as an effective heat sink. The isothermal contours of Figures 7-9 indicate that the components witness a maximum temperature of 356.4 K (83.4˚C), thus there is temperature difference of 73.4˚C in this system. Next, a series of parametric studies was completed each of which is discussed in detail below.

Effect of Thermal Conductivity on Heat Transfer

The primary objective of this current research is to demonstrate the heat spreading characteristics of APG k-core material. Thus, the CFD model described above was exercised by changing the thermal conductivity of the heat sink/cold plate from a very low (bare Aluminum) to a very high value (APG k-core). The results of this parametric study are shown below in Figure 10 which shows the system temperature delta, DT (˚C) plotted as a function of the heat sink/cold plate thermal conductivity k (W/m-K). By inspecting Figure 10 the spreading effects of APG k-core become immediately obvious. The relationship between thermal conductivity and maximum system temperature delta is expected to be linear via the relationship for heat conduction given by the first law of thermodynamics

$$Q = \frac{1}{R_{\text{thermal}}} \Delta T \quad (6)$$

where the thermal resistance for conduction and contact interface is given by

$$R_{\text{thermal}} = R_{\text{contact}} + R_{\text{conduction}} = \frac{1}{h_{\text{contact}}} + \frac{x}{kA} \quad (7)$$

The terms are written in series to indicate that the heat must flow from the heat sink via conduction to the cold plate with an intermediate resistance due to contact in between them. For this study we assumed that the contact h-value for the contact conductance was very large, i.e. since the heat sink/cold plate will ultimately be fabricated as one cohesive unit.

Heat Transfer Coefficient Comparison

One very important piece of information valuable to the electronics cooling community continues to be the elusive film convection heat transfer coefficients. To this end, Figure 11 shows values of convective heat transfer coefficient (W/m^2-K) for a typical CFD simulation of our study. From Figure 11, the convective heat transfer values are found to range from $20 < h < 70$ W/m^2-K for the sub-modules, PCB cards, and component assembly. These values are in agreement with the heat transfer literature [24] for systems of this category.

In order to quantify the effect of inlet fan air speed upon the heat transfer coefficient behavior, a series of with V (ft/min) and L (in). As expected, the handbook correlation of [25] is conservative, i.e. it over predicts the heat transfer coefficient for the range of flow speeds investigated. This is expected, as the uncertainty in a standard textbook based correlation is at least 25% and that the handbook based correlations should not be viewed as sacrosanct [24]. The real mediation to resolve this uncertainty is to perform a system's level test on actual engineering hardware, and then perform a correlation of the numerical model. The data from our CFD study are found to be in qualitative agreement with the simulations and correlations offered by the research of [11,26], which both present plots of Nusselt number versus Reynolds number (i.e. heat transfer coefficient versus flow speed) for heat sinks in electronics cooling applications.

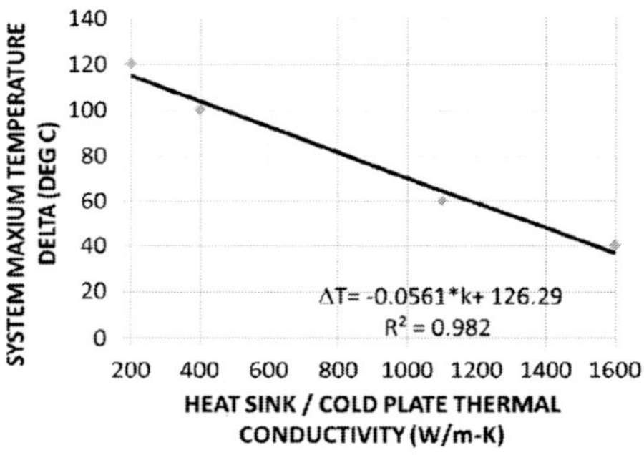

Figure 10: Thermal conductivity effect on heat spreading in CFD simulations.

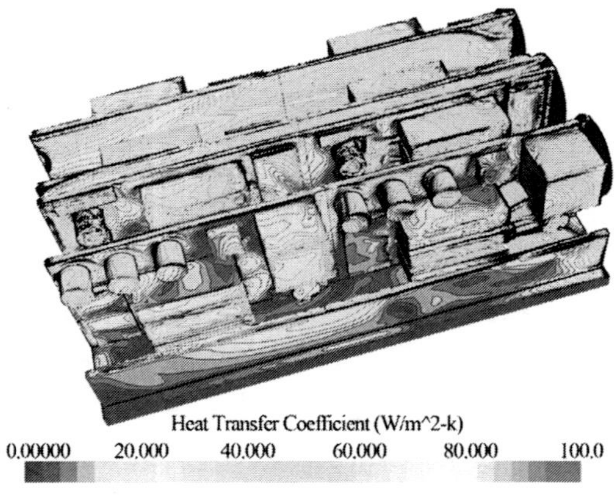

Figure 11: Convective heat transfer coefficient contours.

CFD simulation were performed whereby the flow rate of air was varied, thus resulting in a different average velocity emanating from the fan. Figure 12 shows the results of our CFD simulations in comparison with the laminar flat plate Nusselt number based correlation of [25]. The correlation of [25] is based on the Nusselt number for flat plate laminar flow, and is written as

$$h = 0.00109\sqrt{\frac{V}{L}} \tag{8}$$

Next, the heat transfer coefficient of the system is plotted against temperature delta of the system as shown in Figure 13. Figure 10 shows convective heat transfer coefficient h (W/m^2-K) versus temperature delta (T-T$_0$). The CFD data points are assigned a power-law curve fit and compared to the correlation offered by [23]. The recommended correlation of [23] for electronics cooling is given by

$$h = 2.44\left(\frac{\Delta T}{L}\right)^{1/4} \tag{9}$$

As shown in Figure 13, the hand-book correlation of Equation (9) is found to over-predict the heat transfer coefficient in the range of system temperature differences studied. At this point it should be reiterated that according to [24], the heat transfer correlations given by Equation (8) and Equation (9) are expected to carry an uncertainty of at least 25%. Clearly, the agreement of our current CFD results in Figures 12 and 13 are within the realm of this uncertainty range.

Effect of Cold Plate Coolant Flow Rate

Next we investigate the effects of cold plate coolant flow rate on the system thermal performance. A series of CFD simulations were carried out varying the inlet flow rate of the water in the cold plate from 0.1 kg/sec to 0.2 kg/sec (1.6 gpm to 3.2 gpm for water with r = 1000 kg/m^3). The results are shown in Figure 14 where the heat transfer coefficient of the system is plotted as a function of the flow rate in the cold plate. It should be clarified here that the influence of convective heat transfer coefficient of the entire system, as shown in Figures 7-9, i.e. components and PC boards is plotted in Figure 14, rather than the internal heat transfer coefficient of the tubes comprising the cold plate. As can be seen in Figure 15 the heat exchanger/cold plate assembly maintains a relatively constant temperature displaying only a 5˚C overall spatial gradient. This is primary advantage of utilizing a pumped fluid system cold plate apparatus, i.e. it gives one the ability to hold a somewhat isothermal boundary condition in an actual thermal system.

Effect of Contact Resistance

The next segment of our investigation deals with simulation of the thermal gasketing material comprising the contact conductance interface between the various components in the system. Typical thermal gasket materials used extensively in the commercial electronics cooling industry include Chotherm [26], Thermal Grease [27] and Grafoil [28]. These materials have advertised nominal contact heat transfer coefficients of 5000 W/m^2-K for Chotherm, 50,000 W/m^2-K for Grafoil, and 10,000 W/m^2-K. In practice it is best to apply an area void fraction of at least 50% to these nominal values in order to account for that fact that uniform pressure is not available between the two interfaces being mounted together. This puts a more conservative bounds of the values or thermal contact resistance for these materials as follows, $R = 4 \times 10^{-4}$ K-m^2/W for Chotherm, $R = 4 \times 10^{-5}$ K-m^2/W for Grafoil, and $R = 2 \times 10^{-4}$K-m^2/W for Thermal Grease. Using these values as guidance, CFD simulations were performed on the system level model, whereby all of the interfaces were set to have a contact interference with the values of contact resistance in the range $4 \times 10^{-4} < R < 2 \times 10^{-3}$ K-m^2/W ($2500 < h < 500$ W/m^2-K) thus emulating a "very good" thermal interface to a "very poor" thermal interface. These results are shown in Figure 16 which plots maximum temperature in the system versus thermal contact resistance.

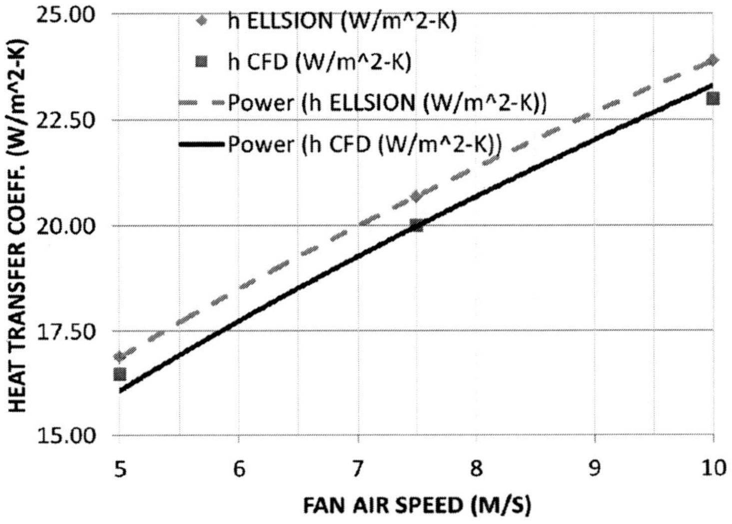

Figure 12: Heat transfer coefficient versus fan air speed.

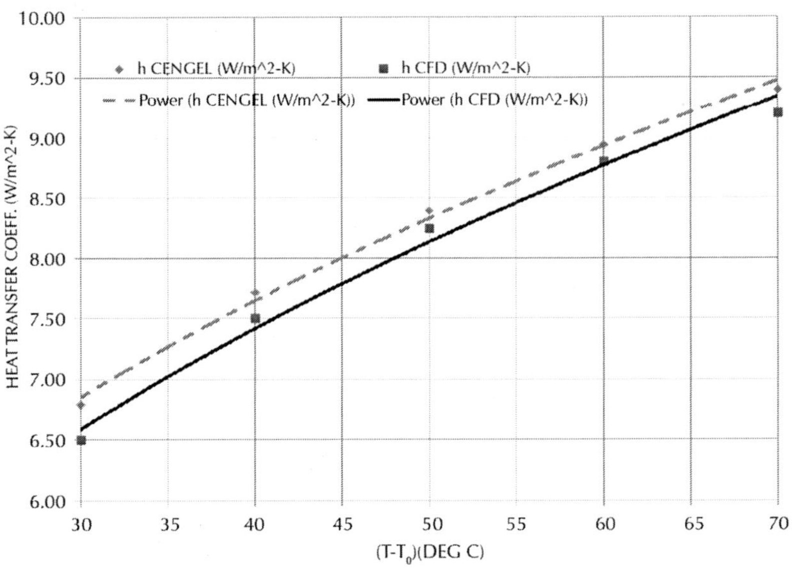

Figure 13: Heat transfer coefficient comparison.

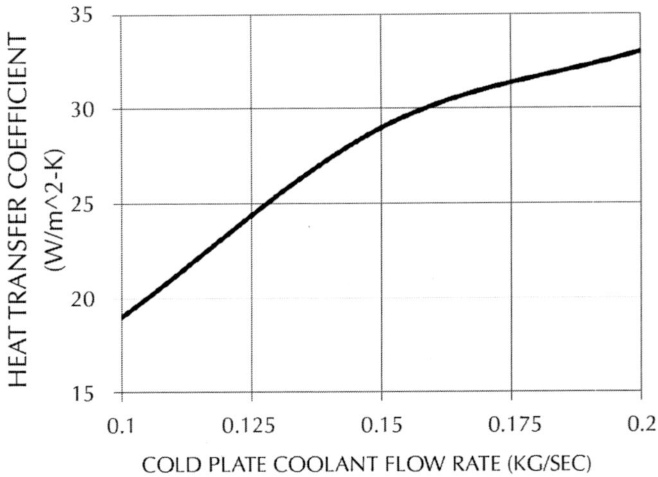

Figure 14: Heat Transfer coefficient versus cold plate flow rate.

As expected the trend of Figure 16 is linear, since the thermal contact resistance is embedded in the conduction heat transfer model of Equation (7) as previously discussed. Figure 17 shows isothermal contours for a submodule which has a very poor thermal contact interface, i.e. $R = 2\times10^{-3}$ K-m^2/K.

Figure 17 shows the range of temperatures, from smallest of 286 K (13°C) where the PCB card is in contact with the heat sink/ cold plate assembly, to a largest value of 350 K (77°C) where the warmest component on the PCB is at the extremity of the PCB card near the air flow region. Here, the impact of a poor thermal contact interface is profound, leading to a 64°C temperature spatial thermal gradient across this particular sub-model.

Thermal Runaway Due to Faulty Components

Thermal runaway due to faulty components in the system are readily simulated using a turn-key CFD model of the sort presented in this paper. Figure 18 shows the effect of simulation of a bad component due to an internal manufacturing flaw(s), an internal short, etc. on a particular PC board. The component experiencing thermal runaway is immediately identified as the outlier in Figure 18, with a case temperature of 580 K (307˚C) while the interface of the sub-modules with the heat sink is 290 K (17˚C), thus a delta of 290˚C is manifested by this single component failure simulation.

Fan Outage Simulations

One very common lead to failure in large data racks and thermally controlled avionics electronics systems is fan failure. This leads to the thermal designer having to carry a redundant system, i.e. back-up fans, in case the primary fan cooling system experiences failure due to power drop out and/or fan motor burn-out. In order to understand the effects of fan failure, various CFD simulations were performed in by varying the air flow around a nominal value of 6 m/s (for a 80 mm dia. fan rated at 60 cfm). Figure 19 shows the results of the fan failure parametric trade study. Figure 19 is a plot of the system level maximum component/chip temperature versus fan air speed.

Figure 15: Heat sink/cold plate temperatures.

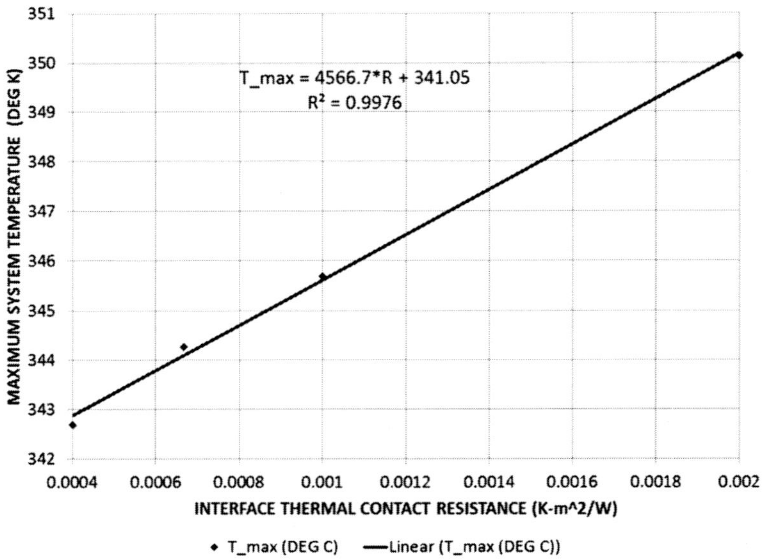

Figure 16: Thermal fault simulation showing effect of contact resistance.

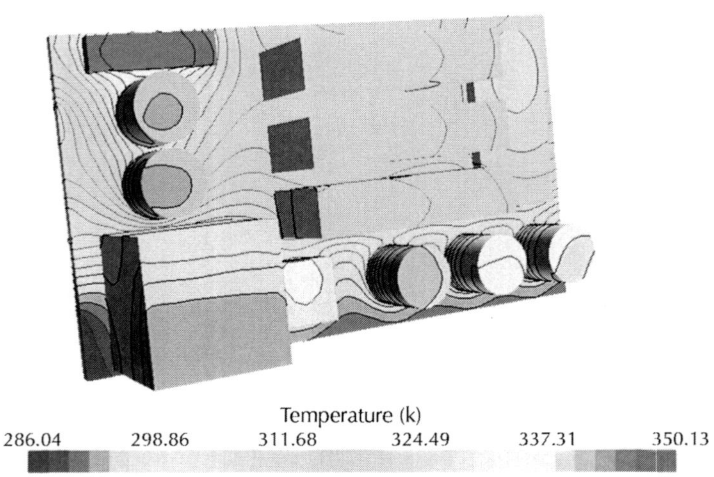

Figure 17: Effect of poor thermal contact resistance.

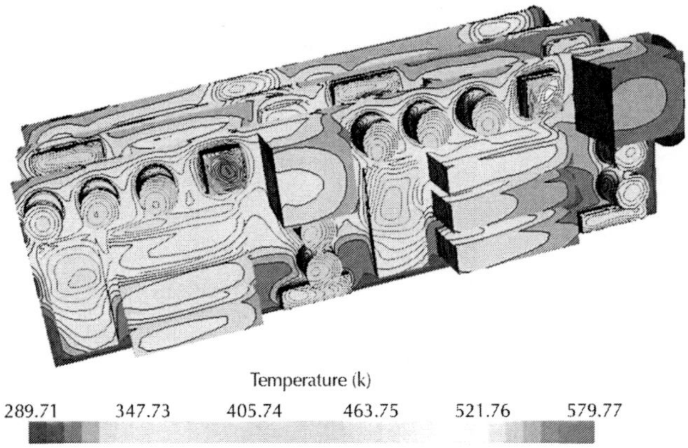

Figure 18: Thermal fault simulation of defective chip.

The trend depicted in Figure 19 is to be expected, i.e. as the fan speed approaches zero, the various components and PC cards within the sub-modules will witness a large temperature. The results of Figure 19 are in agreement with the trends of thermal impedance T/Q versus Reynolds number identified in the study of [29] which focused on the free stream cooling heat transfer cooling of micro-processor chips.

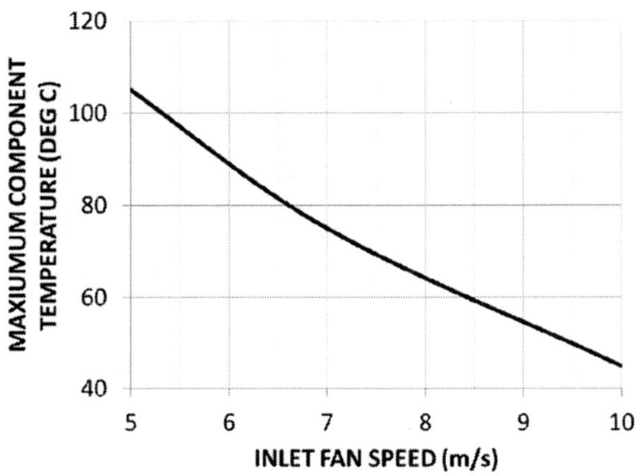

Figure 19: Maximum component/chip temperature versus cooling fan inlet air speed.

CONCLUSIONS

This paper has presented CFD simulations of heat and fluid flow behavior in a moderate sized package of electronics undergoing very large power dissipations on the order of 1000 W/m^2. A combination of heat transfer enhancement thermal control strategies (high thermal conductivity materials, forced convection cooling using DC computer fans, and internal forced convection heat transfer using a pumped fluid loop cold plate/heat exchanger system) have been combined in order to simulate the effectiveness of implementing COTS components to realize a thermal control solution. The results obtained herein are in qualitative agreement with previous studies involving electronics cooling simulations [7, 9, 13, and 16]. Parametric exercises involving air flow rate, water flow rate, thermal interface material, defective electronic components, and failed fans have been carried out.

REFERENCES

1. "Thermacore k-Core Data Sheet," 2013. http://www.thermacore.com/products/kcore.aspx
2. D. D. L. Chung and Y. Takizawa, "Performance of Isotropic and Anisotropic Heat Spreaders," Journal of Electronic Materials, Vol. 41, No. 9, 2012, pp. 2580-2587.
3. Y. S. Yoon, H. Yang and H. Y. Kwak, "Enhancement of the Critical Heat Flux by Using Heat Spreader," KSME International Journal, Vol. 17, No. 7, 2003, pp. 1063- 1072.
4. K. J. Gray, "Effective Thermal Conductivity of a Diamond Coated Heat Spreader," Diamond and Related Materials, Vol. 9, No. 2, 2000, pp. 201-204.
5. P. Mohan, and P. Govindarajan, "Thermal Analysis of CPU with CCC and Copper Base Plate Heat Sinks Using CFD," Heat Transfer Asian Research, Vol. 40, No. 3, 2011, pp. 217-232.

6. A. Part, R. Linton and D. Agonafer, "Coarse and Detailed CFD Modeling of a Finned Heat Sink," IEEE Transactions on Components, Packaging and Manufacturing Technology, Vol. 18, No. 3, 1995, pp. 517-520.
7. P. Mohan and P. Govindarajan, "Experimental and CFD Analysis of Heat Sinks with Base Plate for CPU Cooling," Journal of Mechanical Science and Technology, Vol. 25, No. 8, 2011, pp. 2003-2012.
8. J. Yang, W. Wang and L. Huazhi, "2-Dimensional CFD Simulation and Correlation Development for Optimization of Fin Heatsinks in Electronic Cooling," Journal of Thermal Science, Vol. 10, No. 4, 2001, pp. 363-371.
9. E. Ozturk and I. Tari, "CFD Modeling of Forced Cooling of Computer Chassis," Engineering Applications of Computational Fluid Mechanics, Vol. 1, No. 4, 2007, pp. 304- 313.
10. V. Eveloy, J. Lohan, and P. Rodgers, "A Benchmark Study of Computational Fluid Dynamics Predictive Accuracy for Component-Printed Circuit Board Heat Transfer," IEEE Transactions on Components and Packaging Technologies, Vol. 23, No. 3, 2000, pp. 568-578.
11. I. Tari and Y. Fidan-Seza, "CFD Analyses of a Notebook Computer Thermal Management System and a Proposed Passive Cooling Alternative," IEEE Transactions on Components and Packaging Technologies, Vol. 33, No. 2, 2010, pp. 443-452.
12. A. Part and P. G. Tucker, "CFD Applied to Electronic Systems: A Review," IEEE Transactions on Components, Packaging, and Manufacturing Technology, Vol. 20, No. 4, 1997, pp. 518-529.
13. T. Y. Lee and M. Mahalingam, "Application of a CFD Tool for System Level Thermal Simulation," IEEE Transactions on Components, Packaging and Manufacturing Technology—Part A, Vol. 17, No. 4, 1994, pp. 564-572.

14. T. Y. Lee, B. Chambers and M. Mahalingam, "Application of a CFD Technology to Electronic Thermal Management," IEEE Transactions on Components, Packaging and Manufacturing Technology—Part B," Vol. 18, No. 3, 1995, pp. 511-520. http://dx.doi.org/10.1109/96.404110
15. S. Aradag, U. Olgun, F. Aktuk and B. Baskbuyuk, "CFD Analysis of Cooling Electronic Equipment as an Undergraduate Project," Wiley Periodicals Inc., 2009.
16. J. Choi, Y. Kim, A. Sivasubramaniam, J. Srebric, Q. Wang and J. Lee, "A CFD Based Tool for Studying Temperature in Rack Mounted Servers, IEEE Transactions on Computers," Vol. 57, No. 8, 2008, pp. 1129-1142. http://dx.doi.org/10.1109/TC.2008.52
17. A. Almodi, A. Thompson, N. Kapur, J. Summers, H. Thompson G. and Hannah, "Computational Fluid Dynamics Investigation of Liquid Rack Cooling in Data Centres," Applied Energy, Vol. 89, No. 1, 2012, pp. 150-155.http://dx.doi.org/10.1016/j.apenergy.2011.02.003
18. J. Rambo and Y. Joshi, "Modeling of Data Center Airflow and Heat Transfer: State of the Art and Future Trends," Distributed Parallel Databases, Vol. 21, 2007, pp. 193-225.http://dx.doi.org/10.1007/s10619-006-7007-3
19. R. Boukhanouf and A. Haddad, "A CFD Analysis of an Electronics Cooling Enclosure for Application in Telecommunication Systems," Applied Thermal Engineering, Vol. 30, No. 16, 2010, pp. 2426-2434. http://dx.doi.org/10.1016/j.applthermaleng.2010.06.012
20. Z. Khatir, J. Paton, H. Thompson, N. Kapur, V. Toropov, M. Lawres and D. Kirk, "Computational Fluid Dynamics (CFD) Investigation of Air Flow and Temperature Distribution in a Small Scale Bread-Baking Oven," Applied Energy, Vol. 89, No. 1, 2012, pp. 89-96. http://dx.doi.org/10.1016/j.apenergy.2011.02.002
21. D. B. Flowers and K. R. Anderson, "Numerical Simulation of Conduction Heat Transfer in a System of Slowly Rotating

Concentric Shells Separated by Small Annular Gap Distances," Numerical Heat Transfer Part A, Vol. 46, No. 9, 2004, pp. 1-17.

22. K. R. Anderson, "CFD Analysis of OPALS Sealed Enclosure Electronics Sub-System," Proceedings from the Thermal & Fluids Workshop, NASA JPL, Pasadena, 2012.
23. Y. A. Cengel, "Heat Transfer—A Practical Approach," McGraw-Hill, New York, 2010.
24. F. P. Incropera and D. P. Dewitt, "Heat Transfer," McGraw-Hill, New York, 1991.
25. G. Ellison, "Thermal Computations for Electronics—Conductive, Radiative, and Convective Air Cooling," CRC Press, Boca Raton, 2011.
26. "Graftech Data Sheet," 2013. www.graftech.com
27. D. G. Gilmore, "Spacecraft Thermal Control," Aerospace Press, El Segundo, 2002.
28. "Chotherm Data Sheet," 2013. www.chomerics.com
29. M. A. Ismail, M. Z. Abdullah and M. A. Mujeebu, "A CFD Based Analysis on the Effect of Free Stream Cooling on the Performance of Micro Processor Heat Sinks," International Communications in Heat and Mass Transfer, Vol. 35, No. 6, 2008, pp. 771-778.http://dx.doi.org/10.1016/j.icheatmasstransfer.2008.02.012

Chapter 3

Effects of Thermal Diffusion and Chemical Reaction on MHD Flow of Dusty Visco-Elastic (Walter's Liquid Model-B) Fluid

Om Prakash[1], Devendra Kumar[1], and Y. K. Dwivedi[2]

[1]Department of Mathematics, Hindustan College of Science and technology, Farah Mathura (U.P.)-India.

[2]Department of Mathematics, Ganjdundwara P.G. College Ganjdundwara, Kashiram Nagar (U.P.)-India.

ABSTRACT

The present note consists, the effects of thermal diffusion and chemical reaction on MHD flow of dusty viscous incompressible, electrically conducting fluid between two vertical heated, porous, parallel plates with heat source/sink. The plate temperature is raised linearly with time and concentration level near the plate to C_w. The variable temperature and uniform mass diffusion taking into account the chemical reaction of first order. The series solution method is used to solve the mathematical equations. Effects of various parameters like chemical reaction (K), thermal diffusion (S_T) and magnetic field (M) etc. on velocity profile, skin friction, concentration profile and temperature field are displayed graphically and discussed numerically for different physical parameters. The analysis developed here for thermal diffusion, bears good agreement with real life problems.

INTRODUCTION

Many transport processes exist in nature and industrial application in which the transfer of heat and mass occurs simultaneously as a result of combined buoyancy effects of thermal diffusion and diffusion of chemical species. In the last few decades several efforts have been made to solve the problems on heat and mass transfer in view of their application to astrophysics, geophysics and engineering.

Chemical reaction can be codified either heterogeneous or homogeneous processes. Its effect depends on the nature of the reaction whether the reaction is heterogeneous or homogeneous. A reaction is of order n, if the reaction rate is proportional to the nth power of concentration. In particular, a reaction is of first order, if the rate of reaction is directly proportional to concentration itself. Experimental and theoretical works on MHD flow with thermal diffusion and chemical reaction have been done extensively in various areas i.e. sustain plasma confinement for controlled thermo nuclear fusion, liquid metal cooling of nuclear reactions and electromagnetic casting of metals. Chambre and Yang [1] have worked on thermal diffusion of a chemically reactive species in a laminar boundary layer flow. Dusty viscous and visco-elastic fluids have been discussed by Saffman [2], Micheal and Norey [3], Raptis and Perdikis [4]. Singh [5] proposed the study of free convection and mass diffusion of a dusty visco-elastic (Walter's Liquid Model-B) fluid flowing between two heated porous plates in porous media in presence of magnetic field. References [6,7] focused on the study of convective heat and mass transfer incompressible, viscous Boussinesq fluid in presence of chemical reaction of first order. References [8-10] discussed the effects of thermo diffusion (Soret effects) and diffusion-thermo (Dufour effects) on MHD mixed convection heat and mass transfer of an electrically conducting fluid. Mahantesh et al. [11] studied the boundary layer flow behavior and heat transfer characteristic in Walter's liquid model-B fluid flow. Sharma et al. [12] discussed the unsteady MHD free convection heat and mass transfer of viscous fluid flowing through a Darcian porous regime adjacent to a moving vertical semi-infinite plate under Soret and Dufour effect.

Recently Kumar and Srivastava [13] examined the effects of chemical reaction on MHD flow of dusty viscoelastic (Walters's liquid model-B) liquid with heat source/ sink. The interest of present investigation is to obtain analytical expressions for various profiles like velocity, skin friction for dusty fluid as well as dust particles and also temperature, concentration for dusty fluid. The effects of thermal diffusion parameter (Soret number), chemical reaction parameter etc. are discussed for different profiles.

MATHEMATICAL FORMULATION

We consider the effects of thermal diffusion and chemical reaction on the unsteady dusty flow of an incompressible, slightly conducting, visco-elastic fluid between two heated porous infinite parallel plates (distance 2 h apart) under the influence of uniform magnetic field normal to the flow field in presence of heat source/

sink. We assume x-axis along the flow in the mid-way of the plates and y-axis perpendicular to it. Let u, v be the velocities of dusty fluid and dust particles respectively in the direction of x-axis. The present analysis is based on the following assumptions:

- The flow is in the direction of x-axis and is driven by a constant pressure $\partial p / \partial x$ with negligible body forces.
- The dust particles are non-conducting, solid, spherical, and equal in size, uniformly and symmetrically distributed in the flow field and their number density N_0 is constant throughout the motion.
- There is no externally applied electric field and the induced magnetic field is negligible.
- Initially, when $t \leq 0$, the channel, walls as well as dusty fluid are assumed to be at the same temperature T_0. The foreign mass is assumed to be present at low level and it is uniformly distributed such that it is everywhere C_0.
- When t > 0, the temperature of the walls is instantaneously raised to T_W and the species concentration is raised to C_W.
- There exists a chemical reaction in the mixture.

Under these assumptions and Boussinesq's approximation with concentration, the equations governing the flow are:

$$\frac{\partial u}{\partial t} = g\beta(T - T_0) + g\beta'(C - C_0) +$$

$$\nu(1 - K_0 \frac{\partial}{\partial t})\frac{\partial^2 u}{\partial y^2} + \frac{KN_0}{\rho}(v - u) - \frac{\sigma}{\rho} B_0^2 u - \frac{\nu}{K} u \tag{1}$$

$$m\frac{\partial v}{\partial t} = K'(u - v) \tag{2}$$

$$\frac{\partial T}{\partial t} = \frac{K_T}{\rho C_p}\frac{\partial^2 T}{\partial^2 y} - \frac{Q}{\rho C_p}(T - T_0) \tag{3}$$

$$\frac{\partial C}{\partial t} = D\frac{\partial^2 C}{\partial^2 y} - K_1(C - C_0) + D_T\frac{\partial^2 T}{\partial^2 y} \tag{4}$$

(where the symbols have their usual meaning), at t = 0, the temperature and concentration level changes according to the following laws:

$$T = T_0 + (T_w - T_0)(1 - e^{-at})$$

$$C = C_0 + (C_w - C_0)(1 - e^{-at})$$

The initial and boundary conditions relevant to the problem are:

$$t = 0: u = 0 = v, T = T_0 \quad y \in (-d, d)$$

$$t > 0: u = 0 = v, \quad T = T_0 + (T_w - T_0)(1 - e^{-at}),$$

$$C = C_0 + (C_w - C_0)(1 - e^{-at}) \quad \text{for } y = -d$$

$$u = 0 = v, \quad T = T_0 + (T_w - T_0)(1 - e^{-at}),$$

$$C = C_0 + (C_w - C_0)(1 - e^{-at}) \quad \text{for } y = d \tag{5}$$

We introduce the following non-dimensional quantities,

$$y^* = \frac{y}{d}, u^* = \frac{u}{d}, v^* = \frac{v}{d}, T = \frac{T - T_0}{T_w - T_0},$$

$$C^* = \frac{C - C_0}{C_w - C_0}, t^* = \frac{\nu t}{d^2}, a^* = \frac{d^2 a}{\nu}$$

Introducing these non-dimensional quantities, Equations (1), (2), (3) & (4) reduce to

$$\frac{\partial u}{\partial t} = G_r T + G_m C +$$

$$(1 - E\frac{\partial}{\partial t})\frac{\partial^2 u}{\partial y^2} + \frac{\lambda}{w}(v - u) - \frac{u}{K_1} - M^2 u \tag{6}$$

$$W\frac{\partial v}{\partial t} = (u - v) \tag{7}$$

$$\frac{\partial^2 T}{\partial^2 y} - P_r\frac{\partial T}{\partial t} - ST = 0 \tag{8}$$

$$\frac{\partial^2 C}{\partial^2 y} - S_c\frac{\partial C}{\partial t} - KS_c C + S_T\frac{\partial^2 T}{\partial^2 y} = 0 \tag{9}$$

Now initial and boundary condition (5) according to new system becomeT = 0: u = 0 = v, T = 0 $y \in (-1,1)$

$$\text{T} = 0\text{: u} = 0 = \text{v, T} = 0 \quad y \in (-1,1)$$
$$\text{t} > 0\text{: u} = 0 = \text{v, T} = 1 - \text{e}^{-\text{at}}\text{, C} = 1 - \text{e}^{-\text{at}}\text{, for y} = -1$$
$$\text{u} = 0 = \text{v, T} = 1 - \text{e}^{-\text{at}}\text{, C} = 1 - \text{e}^{-\text{at}}\text{, for y} = 1 \tag{10}$$

where, $\lambda = \frac{mN_0}{\rho}$ (mass concentration of dust particle)

M (Hartmann number) $= B_0 d\sqrt{\frac{\sigma}{\mu}}$

W (relaxation time parameter for particles) $= \frac{mv}{K'd^2}$

G_r (grashof number) $= \frac{g\beta d(T_w - T_0)}{v}$

G_m (Modified grashof number) $= \frac{g\beta' d(C_w - C_0)}{v}$

S_c (Schmidt number) $= \frac{\nu}{D}$

S (heat source/sink parameter) $= Q\frac{d^2}{K_T}$

K (Chemical reaction parameter) $= \frac{K_1 d^2}{\nu}$

$E = \frac{K_0 \nu}{d^2}$ (Visco-elastic parameter), $K_1 = \frac{k}{d^2}$, P_r (Prandtl number) $= \frac{\mu C_P}{K_T}$

$S_T = \frac{D_T(T_w - T_0)}{D(C_w - C_0)}$ (Thermal diffusion parameter)

METHOD OF SOLUTION

To solve the Equations (6) to (9) subject to the boundary conditions (10), according to I. Pop [14], we assume

$$u(y,t) = u_0(y) + \varepsilon u_1(y,t)e^{-at} + \ldots\ldots\ldots\ldots\ldots$$

$$v(y,t) = v_0(y) + \varepsilon v_1(y,t)e^{-at} + \ldots\ldots\ldots\ldots\ldots$$

$$T(y,t) = T_0(y) + \varepsilon T_1(y,t)e^{-at} + \ldots\ldots\ldots$$

$$C(y,t) = C_0(y) + \varepsilon C_1(y,t)e^{-at} + \ldots\ldots\ldots\ldots\ldots \quad (11)$$

Substituting the equations like (11) into the Equations (6) to (9) and equating harmonic and non-harmonic terms, we get the following set of equations.

$$u_0'' - \left(\frac{1}{K_1} + M_1^2\right)u_0 = -G_r T_0 - G_m C_0 \quad (12)$$

$$(1+aE)u_1'' + \frac{\lambda}{w}u_1 -$$

$$(\frac{1}{K_1} + M_1^{\ 2} - a)u_1 - \frac{\lambda}{w}v_1 = -G_r T_1 - G_m C_1 \tag{13}$$

$$u_0 = v_0 \ \& \ u_1 = v_1 \ (1\text{-aw}) \tag{14}$$

$$T_0'' - ST_0 = 0 \tag{15}$$

$$T_1'' - (S\text{-aPr})\, T_1 = 0 \tag{16}$$

$$C_0'' - KS_c C_0 + S_T T_0'' = 0 \tag{17}$$

$$C_1'' - (K-a)S_c C_1 + S_T T_1'' = 0 \tag{18}$$

Where dashes represents differentiation w. r. to y.

Boundary conditions are reduced to:

$$u_0 = v_0 = u_1 = v_1,\ T_0 = C_0 = 1,$$

$$T_1 = C_1 = -\frac{1}{\varepsilon},\ at\ y = -1$$

$$u_0 = v_0 = u_1 = v_1,\ T_0 = C_0 = 1,$$

$$T_1 = C_1 = -\frac{1}{\varepsilon},\ at\ y = 1 \tag{19}$$

Solutions of the Equations (12) to (18) under the boundary conditions (19) after substituting in (11), we have:

$$u(y,t) = (A_{11} + A_{12})\frac{CoshM_1 y}{CoshM_1} - A_{11}\frac{CoshAy}{CoshA} - A_{12}\frac{CoshA_0 y}{CoshA_0} + \left\{-(A_{13} + A_{14})\frac{CoshM_2 y}{CoshM_2} + A_{13}\frac{CoshA_1 y}{CoshA_1} + A_{14}\frac{CoshA_2 y}{CoshA_2}\right\}e^{-at}$$

$$V(y,t) =$$

$$(A_{11} + A_{12})\frac{CoshM_1 y}{CoshM_1} - A_{11}\frac{CoshAy}{CoshA} - A_{12}\frac{CoshA_0 y}{CoshA_0} + \left(\frac{1}{1-aw}\right)\left\{-(A_{13} + A_{14})\frac{CoshM_2 y}{CoshM_2} + A_{13}\frac{CoshA_1 y}{CoshA_1} + A_{14}\frac{CoshA_2 y}{CoshA_2}\right\}e^{-at}$$

$$C(y,t) = (1 - A_m)\frac{CoshA_0 y}{CoshA_0} + A_m \frac{CoshAy}{CoshA} + \left\{(A_n - 1)\frac{CoshA_2 y}{CoshA_2} - A_n \frac{CoshA_1 y}{CoshA_1}\right\}e^{-at}$$

$$T(y,t) = \frac{CoshAy}{CoshA} - \frac{CoshA_1 y}{CoshA_1}e^{-at}$$

Skin Friction

Let τ_f and τ_P be the skin friction for dusty fluid and dust particles respectively then we have:

$$\tau_p = \left|\frac{\partial v}{\partial y}\right|_{y=1} = (A_{11} + A_{12})M_1 \tanh M_1 - A_{11} A \tanh A -$$

$$A_{12} A_0 \tanh A_0 + \left(\frac{1}{1 - aw}\right)[-(A_{13} + A_{14})M_2 \tanh M_2 +$$

$$A_{13} A_1 tanhA_1 + A_{14} A_2 \tanh A_2]e^{-at}$$

$$\tau_f = \left|\frac{\partial u}{\partial y}\right|_{y=1} = (A_{11} + A_{12}).M_1 \tanh M_1 - A_{11} A \tanh A -$$

$$A_{12} A_0 \tanh A_0 + [-(A_{13} + A_{14})M_2 \tanh M_2 +$$

$$A_{13} A_1 tanhA_1 + A_{14} A_2 \tanh A_2]e^{-at}$$

RESULTS & DISCUSSION

Numerical solutions for velocity profile, skin friction for dusty fluid as well as dust particles and also temperature field, concentration profile for dusty fluid have been calculated. The values of different parameters and their effects on velocity, Temperature, concentration and skin friction have been displayed through graphs.

A temperature field has been represented in Figure 1, which indicates the effects of heat source/sink parameter and Prandtl number.

Temperature Field for Different Values of S and P_r (t = 1, a = 0.2)

It is observed that increasing values of heat source/sink parameter and Prandtl number decreases the temperature. Also we see that the temperature is minimum at the centre of the channel (y = 0) and increasing towards the plates.

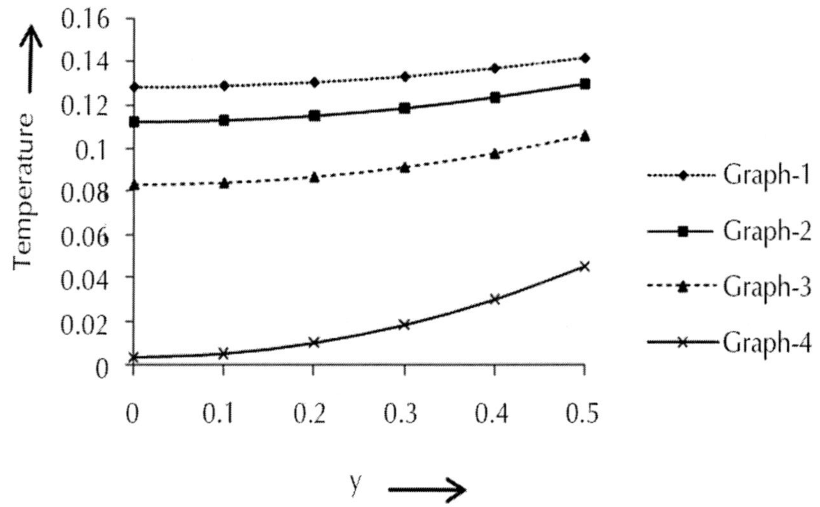

Figure 1: Temperature filed for different values of S and P_r.

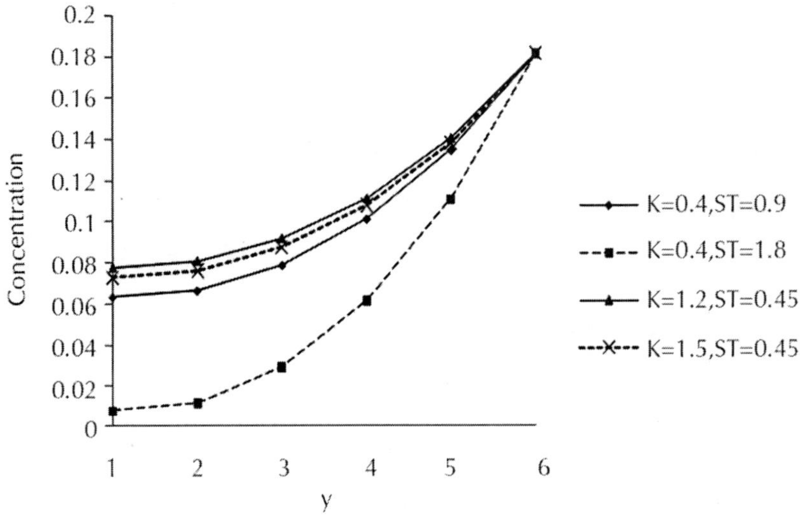

Figure 2: Concentration profile for S_T.

Concentration Profile for the Different Values of K and S_T ($a = 0.2$, $S = 0.2$, $t = 1$, $P_r = 0.71$, $S_c = 0.6$, $D = 1$)

Observation of Figure 2 is increasing value of thermal diffusion parameter (Soret number) and chemical reaction parameter decreases the concentration. Concentration is minimum at the centre of the channel ($y = 0$) and increasing towards the plates.

Velocity Profile for the Different Values of K and S_T ($a = 0.2$, $S = 0.2$, $t = 1$, $P_r = 0.71$, $S_c = 2$, $D = 1$, $w = 0.5$, $E = 1$, $K_1 = 10$, $G_m = 5$, $G_r = 10$)

From Figures 3 and 4 we observe that increasing value of thermal diffusion parameter (Soret number) increases the velocity of dusty fluid and dust particles while chemical reaction parameter decreases the same. Also Figures 3 and 4 bears that the velocity is maximum at the centre of the channel and decreasing towards the plates.

Skin Friction Profile for Fluid and Dust Particles: (a = 0.2, S = 0.2, t = 1, P_r= 0.71, S_c = 2, D = 1, w = 0.5, E = 1, K_1 = 10, G_m = 5, G_r = 10)

Skin friction for different values of S_T (M = 3, K = 0.4, S = 0.3):

Skin friction for different values of S, (M = 3, K = 1, S_T = 1.4):

Skin friction for different values of K (M = 3, S_T = 2.7.4, S = 0.3):

Skin friction for different values of M (S_T = 1.4, K = 1, S = 0.4):

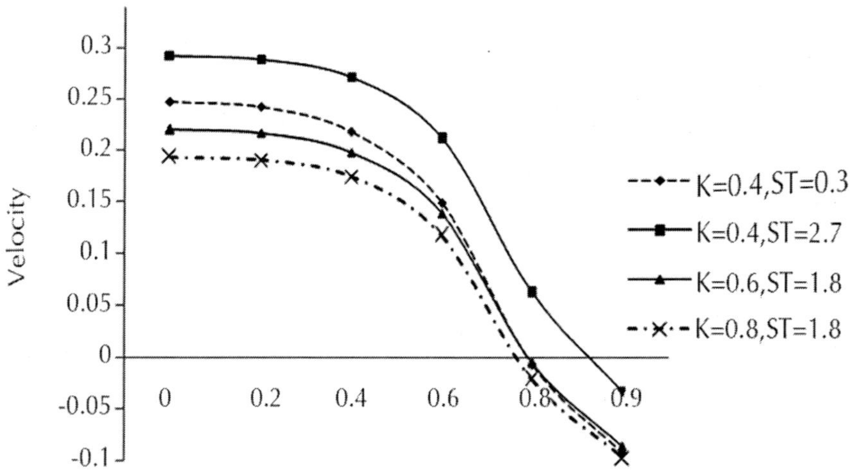

Figure 3: Velocity profile for dusty fluid.

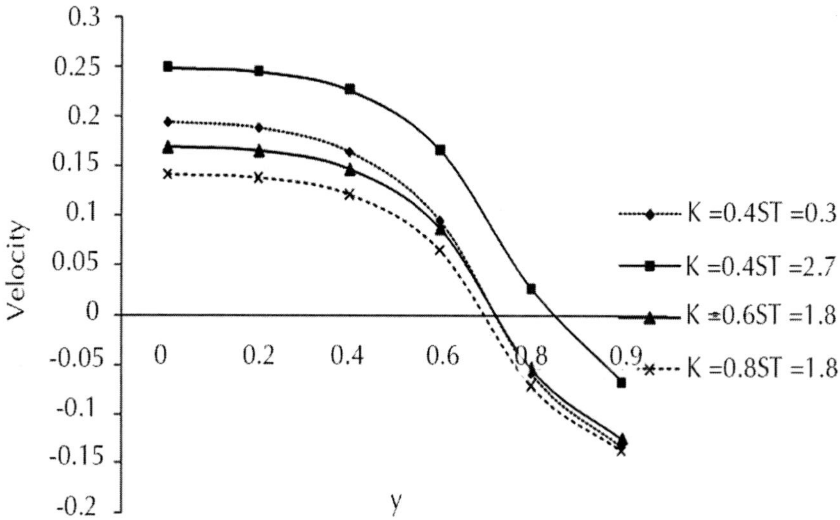

Figure 4: Velocity profile for dust particle.

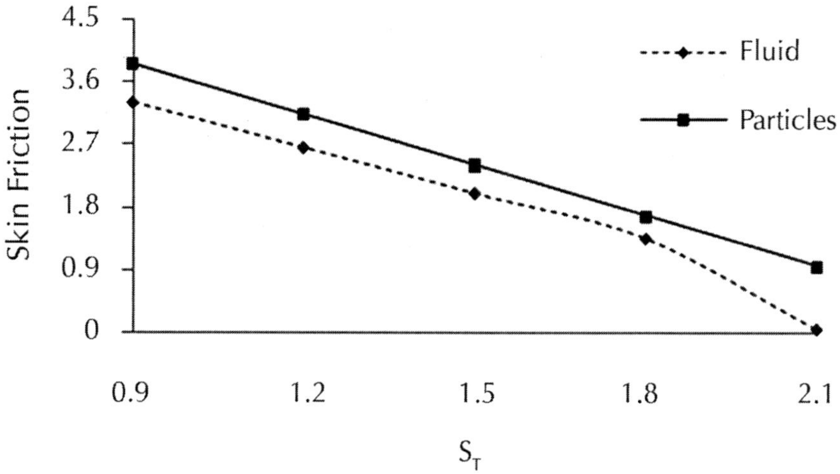

Figure 5.1: Skin friction for S_T.

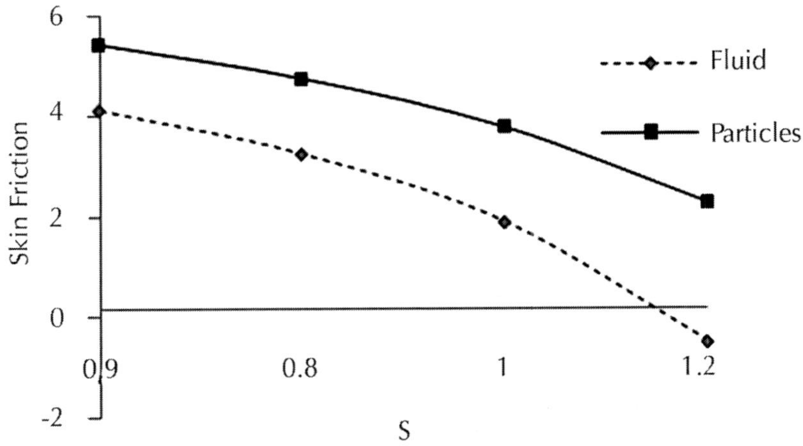

Figure 5.2: Skin friction for S.

The results displayed in Figures 5.1-5.4 are as, the increasing value of thermal diffusion parameter and heat source/sink parameter decreases the skin friction of dusty fluid and dust particles. Increasing value of chemical reaction parameter and magnetic field parameter increases the skin friction of dusty fluid and dust particles.

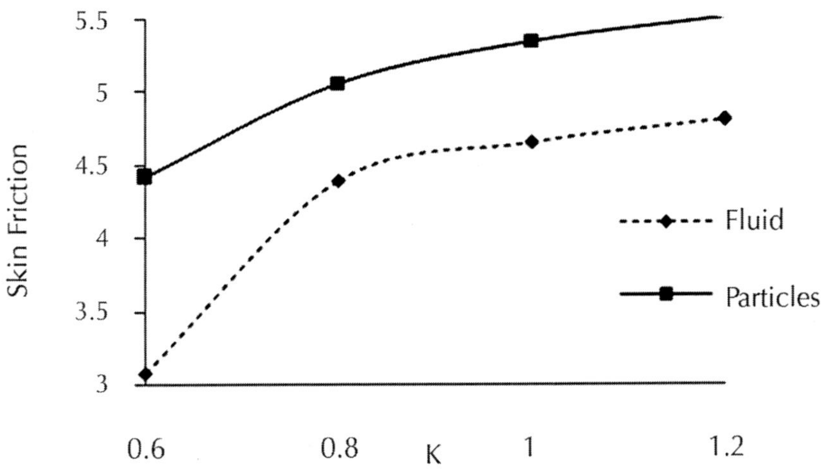

Figure 5.3: Skin friction for K.

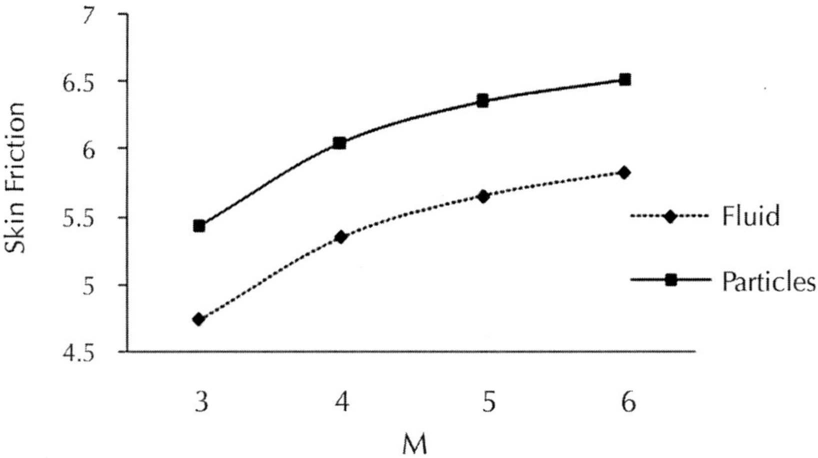

Figure 5.4: Skin friction for M.

CONCLUSIONS

The theoretical and numerical solutions are obtained for different profiles. From graphical representations, we have the following observations:

- Velocity and skin friction of the dust particles behaves same as dusty fluid.
- Increasing value of y increases the temperature, concentration while decreases the velocity of dusty fluid and dust particles.
- Velocity of dust particles is less than velocity of dusty fluid and skin friction of dust particles is greater than that of dusty fluid.
- Increasing values of thermal diffusion parameter (Soret number) decreases the concentration, skin friction while increases the velocity of dust particles and dusty fluid.
- Increasing values of magnetic field parameter increases the skin friction for both the dusty fluid and dust particles.

- Increasing values of heat source/sink parameter decreases the skin friction for the dusty fluid as well as dust particles.

ACKNOWLEDGMENTS

The author is sincerely thankful to the reviewers for the critical comments and suggestions to improve the quality of the manuscript.

REFERENCES

1. P. L. Chambre and J. D. Young, "On Diffusion of a Chemically Reactive Species in a Laminar Boundary Layer Flow," Physics of fluids, Vol. 1, 1958, pp. 40-54.
2. P. G. Saffman, "On the Stability of Laminar Flow of a Dusty Gas," Journal of fluid mechanics, Vol. 13, No. 1, 1962, pp. 120-129.
3. D. H. Michael and P. W. Norey, "The Laminar Flow of a Dusty Gas between Rotating Cylinders," The Quarterly Journal of Mechanics and Applied Mathematics, Vol. 21, 1968, pp. 375-388.
4. A. Raptis and C. P. Perdikis, "Oscillatory Flow through a Porous Medium by the Presence of Free Convective Flow," International Journal of Engineering Science, Vol. 23, 1985, pp. 51-55.
5. N. P. Singh, A. K. Singh, M. K. Yadav and A. K. Singh, Acta Ciencia Indica. Vol. XXVIII M, No. 1, 2002, pp. 089.
6. I. U. Mbeledogu and A. Ogulu, "Heat and Mass Transfer of an Unsteady MHD Natural Convection Flow of a Rotating Fluid Past a Vertical Porous Flat Plate in the Presence of Radiative Heat Transfer," International Journal of Heat and Mass Transfer, Vol. 50, No. 9-10, 2007, pp. 1902-1908.
7. P. M. Patil and P. S. Kulkarni, "Effects of Chemical Reaction on Free Convective Flow of a Polar Fluid through a Porous Medium in the Presence of Internal Heat Generation,"

International Journal of Thermal Sciences, Vol. 47, No. 8, 2008, pp. 1043-1054.

8. E. Osalusi, J. Side and R. Harris, "Thermal Diffusion and Thermo Effect on Combined Heat and Mass Transfer of a Steady MHD Convective and Slip Flow due to a Rotating Disk with Viscous Dissipation and Ohmic Heating," International Communications in Heat and Mass Transfer, Vol. 35, No. 8, 2008, pp. 908-915.
9. A. A. Afify, "Similarity Solution in MHD: Effects of Thermal Diffusion and Diffusion Thermo on Free Convective Heat and Mass Transfer over a Stretching Surface Considering Suction or Injection," Communications in Nonlinear Science and Numerical Simulation, Vol. 14, No. 5, 2009, pp. 2202-2214.
10. O. A. Bég, A. Y. Bakier and V. R. Prasad, "Numerical Study of Free Convection Magnetohydrodynamic Heat and Mass Transfer from a Stretching Surface to a Saturated Porous Medium with Soret and Dufour Effects," Computational Materials Science, Vol. 46, No. 1, 2009, pp. 57-65.
11. M. M. Nandeppanavar, M. S. Abel and J. Tawade, "Heat Transfer in a Walter's Liquid B Fluid over an Impermeable Stretching Sheet with Non-Uniform Heat Source/ Sink and Elastic Deformation", Communications in Nonlinear Science and Numerical Simulation, Vol. 15, No. 7, 2010, pp. 1791-1802.
12. R. Sharma, R. Bhargava and P. Bhargava, "A Numerical Solution of Steady MHD Convection Heat and Mass Transfer on a Semi Infinite Vertical Porous Moving Plate Using Element Free Galerkin Method." Computational Materials Science, Vol. 48, No. 3, 2010, pp. 537-543.
13. D. Kumar and R. K. Srivastava, "Effects of Chemical Reaction on MHD Flow of Dusty Visco-Elastic (Walter's Liquid model-B) Liquid with Heat Source/Sink," Proceeding of National Seminar on Mathematics and Computer Science, Meerut, 2005, pp. 105-112.
14. I. Pop, Rev. Roum. Physics, Vol. 13, 1968, pp. 41.

Chapter

4

Investigation of a Counter Flow Microchannel Heat Exchanger Performance with Using Nanofluid as a Coolant

Mushtaq I. Hasan[1], Abdul Muhsin A. Rageb[2], and Mahmmod Yaghoubi[3]

[1]Department of Mechanical Engineering, College of Engineering, Thi-Qar University, Nasiriyah, Iraq

[2]Department of Mechanical Engineering, College of Engineering, Basra University, Basra, Iraq

[3]Academy of Sciences of Iran, Tehran, Iran

ABSTRACT

In this paper the performance of a counter flow microchannel heat exchanger (CFMCHE) is numerically investigated with a nanofluid as a cooling medium. Two types of nanofluids are used Cu-water and Al_2O_3-water. From the results obtained it's found that thermal performance of CFMCHE increased with using the nanofluids as cooling medium with no extra increase in pressure drop due to the ultra-fine solid particles and low volume fraction concentrations. The nanofluids (Cu-water and Al_2O_3-water) volume fractions were in the range 1% to 5%. It's also found that nanofluid-cooled CFMCHE could absorb more heat than water-cooled CFMCHE when the flow rate was low. For high flow rates the heat transfer was dominated by the volume flow rate and nanoparticles did not contribute to the extra heat absorption. Also the performance of CFMCHE can be increased considerably by using nanofluids with higher thermal conductivities.

INTRODUCTION

One of the methods for enhancing heat transfer is the application of additives to the working fluid. The basic idea is to enhance heat transfer by changing the fluid transport properties such as in nanofluid where the solid particles added to the base fluid to increase its thermal conductivity. Nanofluid consists of a base fluid such as water and nano scale metallic or non-metallic particles. The term of nanofluids was first used by the group at the Argonne National Laboratory, USA, about a decade ago. Then using a nanofluids as a heat transfer working fluid has gained much attention in recent years due to its potential advantages include higher thermal conductivity than the pure fluids, excellent stability and little increase in pressure drop.

Tuckerman and Pease [1] introduced the idea of using microchannel heat sink (MCHS) for cooling of electronic devices in 1981. They experimentally demonstrated the MCHS capability and claimed that they were able to dissipate heat flux rate of 790 W/ cm^2. They showed that the convective heat transfer of single phase flows could be improved by decreasing the width of the heat sink channels and by increasing the area wetted by the heat transfer fluid. The experimental and analytical studies by Wang et al. [2], Lee et al. [3], Wang et al. [4] and Koo and Kleinstreuer [5] showed that nanofluids have higher thermal conductivity than pure fluids and therefore it has great potential for heat transfer enhancement.

Li and Xuan [6], Xuan and Li [7] and Pak and Cho [8] experimentally measured the convection heat transfer and pressure drop for nanofluid tube flows. Their results indicated that heat transfer coefficient was greatly enhanced and depends on the Reynolds number, the particle size and shape, and particle volume fraction. They also found that nano particles did not cause an extra pressure drop.

Donsheng and Yulog [9] studied experimentally the convective heat transfer of nanofluid made of -Al_2O_3- water, flowing through a copper tube in the laminar flow regime. They showed a considerable enhancement of convective heat transfer using the nanofluids. The enhancement was particularly significant in the entrance region, and was much higher than that solely due to the enhancement on thermal conduction.

Seok and Choi [10] investigated numerically the cooling performance of microchannel heat sink with nanofluids. They showed that the cooling performance of a MCHS with water based nanofluids containing diamond (1% volume fraction and 2 nm) at the fixed pumping power of 2.25 W is enhanced by about 10% compared with that of a MCHS with water.

Tsung and Reigu [11] studied analytically the performance of microchannel heat sink (MCHS) using copper-water (Cu-H_2O) and carbon nanotube-water (CNTH_2O) nanofluids as coolants. Velocity and temperature distributions were obtained by modeling the MCHS as a porous media. The resulting velocity and temperature were then used to evaluate the thermal resistance that characterizes MCHS performance. They found that the nanofluid reduces the temperature difference between the MCHS bottom wall and bulk nanofluid compared with that from pure fluid, which lead to produces a reduction in conductive thermal resistance.

Joescon and Issam [12] performed experiments to explore the microchannel cooling benefits of Al_2O_3-water nanofluid. They found that the high thermal conductivity of nano particles enhance the single phase heat transfer coefficient especially for laminar flow. Higher heat transfer coefficient was achieved mostly in the entrance region of the microchannels and the enhancement was weaker in the fully developed region, providing that nano particles have an appreciable effect on thermal boundary layer development. It was also observed that higher concentrations also produced greater sensitivity to heat flux.

Mushtaq et al. [13] investigated numerically the effect of channels geometry (the size and shape of channels) on the performance of counter flow microchannel heat exchanger and used liquid water as a cooling fluid. They found that with decreasing the size of channels both the effectiveness of heat exchanger and pressure drop were increased and they claimed that the decision of increaseing or decreasing the size of channels depends on the application in which this heat exchanger is used. Also they found that the circle is the best shape for the channels of this type of heat exchangers since it gives higher overall performance (including both the hydrodynamic and thermal performance).

Mushtaq I. Hasan [14] numerically investigated the performance of counter flow microchannel heat exchanger with MEPCM suspension as a cooling fluid. He fund that, using of MEPCM suspension lead to improve thermal performance of CFMCHE but it also lead to extra increase in pressure drop and as results the overall performance decreased with using this suspension as a cooling medium.

Most of the studies made to investigate heat transfer and flow characteristics of nanofluids were carried out in macro-scale dimensions. Only a few studies addressed the nanofluid flow and heat transfer in micro-scale dimensions [11]. In this paper an investigation is made to study the effect of using nanofluids as a cooling medium on the performance of CFMCHE.

MATHEMATICAL MODEL

For modeling, nanofluid is treated as a single-phase fluid. This assumption can be applicable since the particles are ultra-fine and they are easily fluidized [6, 7]. Moreover, the particle volume fraction in nanofluid is usually low.

Under such assumptions the governing equations for the nanofluid flow and heat transfer are greatly simplified and local fluid and particles are in thermal equilibrium.

Schematic structure of the studied counter flow microchannel heat exchanger with square channels is shown in Figure 1. To study the entire CFMCHE numerically it is complicated and needs huge of CPU time. Therefore and due to the geometrical and thermal symmetry between hot and cold channels rows, an individual heat exchange unit which consists of two channels containing hot and cold fluids and a separating wall is considered as shown in Figure 2 will be used as a model to represent the complete counter flow microchannel heat exchanger since it give an adequate indication about its performance and the heat is transferred from hot fluid to cold fluid through the thick wall separating them.

The governing equations for 3D, laminar, steady and incompressible with constant fluid properties are:

$$\nabla V_j = 0 \tag{1}$$

$$\rho_j \left(V_j \cdot \nabla V_j \right) = -\nabla P_j + \nabla \cdot \left(\mu_j \nabla V_j \right) \tag{2}$$

$$\rho_j C_{pj} \left(V_j \cdot \nabla T_j \right) = k_j \nabla^2 T_j \tag{3}$$

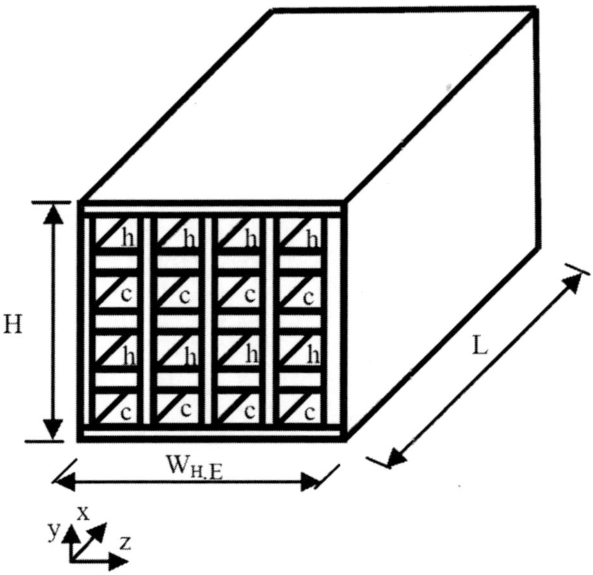

Figure 1: A schematic model of the counter flow MCHE.

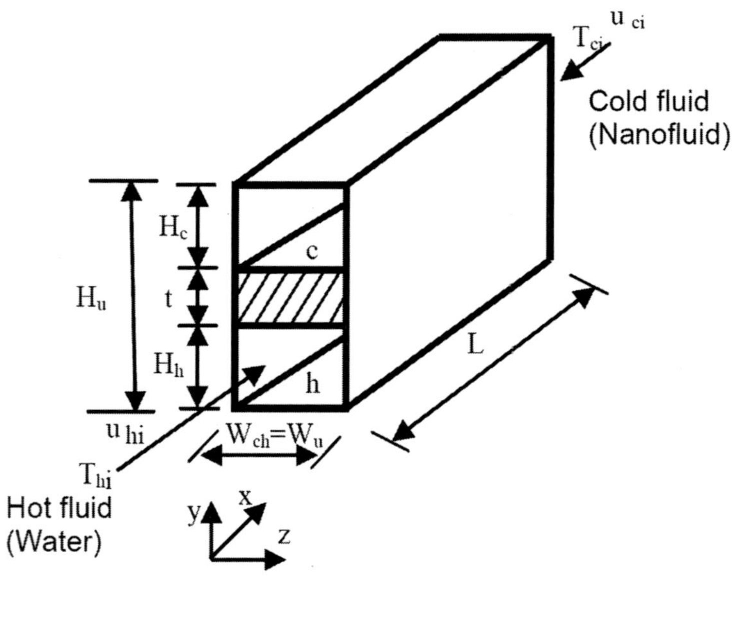

Figure 2: A schematic of heat exchange unit.

For solid

$$k_s \nabla^2 T_s = 0 \tag{4}$$

where j = h and c for hot and cold channels respectively.

The boundary conditions used are:

For lower channels (hot fluid) (water) (0 ≤ y ≤ Hh)

Location	B.C	Comments
x = 0	uh = uhi, vh = wh = 0 Th = Thi	Hot fluid inflow
x = L	$\frac{\partial u_h}{\partial x} = v_h = w_h = 0$	Hot fluid outflow (fully developed flow end of channel)
	$\frac{\partial T_h}{\partial x} = 0$	
	uh = vh = wh = 0	
z = 0	$\frac{\partial T_h}{\partial x} = 0$	No-slip, adiabatic wall
	uh = vh = wh = 0	
z = Wch	$\frac{\partial T_h}{\partial x} = 0$	No-slip. adiabatic wall
	uh = vh = wh = 0	
y = 0	$\frac{\partial T_h}{\partial x} = 0$	No-slip. adiabatic wall
	uh = vh = wh = 0	
y= Hh	$-k_h \frac{\partial T_h}{\partial y} = -k_s \frac{\partial T_s}{\partial y}$	Fluid-solid interface(no-slip, conjugate heat transfer)
	Th=Ts	

For upper channel (cold fluid) (Nanofluid) (Hh + t ≤ y ≤ Hh + t + Hc)

Location	B.C	Comments
x = 0	$\frac{\partial u_c}{\partial x} = v_c = w_c = 0$	Cold fluid inflow(fully developed flow end of channel)
	$\frac{\partial T_c}{\partial x} = 0$	
x = L	uc=uci,vc=wc=0,	Cold fluid inflow
	Tc = Tci	
	uc=vc=wc =0	
z = 0	$\frac{\partial T_c}{\partial z} = 0$	No-slip, adiabatic wall
	uc =v=wc=0	
z = Wch	$\frac{\partial T_c}{\partial z} = 0$	No-slip. adiabatic wall
	uc = vc = wc =0	
y =Hh+t	$-k_c \frac{\partial T_c}{\partial y} = -k_s \frac{\partial T_s}{\partial y}$	Fluid-solid interface(no-slip, conjugate heat transfer)
	Tc=Ts	
	uc =vc=wc=0	
y= Hh+tHc	$\frac{\partial T_c}{\partial y} = 0$	No-slip. adiabatic wall

For solid wall separating two channels (Hh ≤ y ≤ Hh + t)

Location	B.C	Comments
x = 0	$\frac{\partial T_s}{\partial x} = 0$	Adiabatic wall
x = L	$\frac{\partial T_s}{\partial x} = 0$	Adiabatic wall
z = 0	$\frac{\partial T_s}{\partial z} = 0$	Adiabatic wall
z = Wch	$\frac{\partial T_s}{\partial z} = 0$	Adiabatic wall
y =Hh	$-k_h \frac{\partial T_h}{\partial y} = -k_s \frac{\partial T_s}{\partial y}$	Fluid-solid interface
	Th=Ts	
y= Hh+t	$-k_c \frac{\partial T_c}{\partial y} = -k_s \frac{\partial T_s}{\partial y}$	Fluid-solid interface
	Tc=Ts	

By solving the above governing equations the distribution of velocity, pressure and temperature are determined in the fluids and solid domains. From these distributions the other parameters such as effectiveness, heat transfer rate, pressure drop, pumping power required and overall performance index are determined.

Heat exchanger effectiveness is the ratio of the actual heat transfer to the maximum possible heat that can be transferred:

$$\varepsilon = \frac{q}{q_{\max}} \quad (5)$$

where

$$q_{\max} = C_{\min}\left(T_{hi} - T_{ci}\right) \tag{6}$$

and

$$q = C_h\left(T_{hi} - T_{ho}\right) = C_c\left(T_{co} - T_{ci}\right)$$

where

$$C_h = \dot{m}Cp_h$$

and

$$C_c = \dot{m}Cp_c$$

$$\varepsilon = \frac{C_h\left(T_{hi} - T_{ho}\right)}{C_{\min}\left(T_{hi} - T_{ci}\right)} = \frac{C_c\left(T_{co} - T_{ci}\right)}{C_{\min}\left(T_{hi} - T_{ci}\right)} \tag{7}$$

The pumping power required to circulate the hot and cold fluids in CFMCHE is:

$$PP = \dot{V}\Delta Pt \tag{8}$$

where $\dot{V}$ is the volumetric flow rate (m^3/s) and Pt is total pressure drop in heat exchange unit.

$$\dot{V} = v_{in}A \tag{9}$$

$$\Delta P_t = \Delta P_h + \Delta P_c = \left(P_{hi} - P_{ho}\right) + \left(P_{ci} - P_{co}\right) \tag{10}$$

To calculate the overall performance of the CFMCHE taking into consideration both thermal and hydrodynamic performances, the so called performance index is used which is the ratio of CFMCHE effectiveness to the total pressure drop is used [13]:

$$\eta = \frac{\varepsilon}{\Delta Pt} \tag{11}$$

Other factor used to calculate the overall performance and to verify the results of performance index which is defined as the ratio of heat transfer rate to the pumping power required $\left(\frac{q(W)}{PP(W)}\right)$ [13, 14].

PROPERTIES OF NANOFLUIDS

The thermo physical properties of the nanofluids are mainly depend upon the properties of the base fluid and the solid particles, volume fraction of the solid particles in the suspension and particles shape. The properties of nanofluids can be calculated using the following relations [11, 12]:

Thermal conductivity:

$$k_{nf} = k_f \left[\frac{k_p + (SH-1)k_f - (SH-1)c(k_f - k_p)}{k_p + (SH-1)k_f + c(k_f - k_p)} \right] \quad (12)$$

Viscosity:

$$\mu_{nf} = \mu_f (1 + 2.5c) \quad (13)$$

Density:

$$\rho_{nf} = c\rho_p + (1-c)\rho_f \quad (14)$$

Specific heat:

$$Cp_{nf} = cCp_p + (1-c)Cp_f \quad (15)$$

where SH is solid particle shape factor.

$$SH = \frac{3}{\psi} \quad (16)$$

ψ Is sphericity defined as the ratio of the surface area of a sphere with a volume equal to that of the particle to the surface area of the particle. For spherical particles SH = 3.

k_f, k_{nf}, k_p are thermal conductivities of the base fluid, nanofluid and solid particles respectively.

In this paper two nanofluids were used. Water is the base fluid used for both of them and the solid particles are cupper (Cu) for the first nanofluid and Aluminum oxide (Al_2O_3) for the second. So the two nanofluids used are (Cu-water) and (Al_2O_3-water) and the volume fraction used are (1%, 2%, 3%, 4%, 5%). The relation for viscosity, equation (13) is used up to c = 5% [11].

The properties of two nanofluids used (Cu-water) and (Al_2O_3-water) are listed in Tables 1 and 2 [15].

NUMERICAL SOLUTION

A computational fluid dynamic code FLUENT 6.3 is used to solve the governing equations and calculate the distributions of the flow velocity, pressure and temperature in a CFMCHE. For square channel with length L = 10 mm, channel height H = 100 μm, width W = 100 μm, wall thickness t = 50 μm, the mesh used is hexahedral elements. For a flow of Re = 50, three mesh sizes were used. The first mesh set was (16 × 16 × 90) and (16 × 8 × 90) in z, y, x directions for channels and wall respectively. The second mesh was (20 × 20 × 110), and (20 × 10 × 110) and the third mesh was (25 × 25 × 110) and (25 × 15 × 110). The computational results for CFMCHE effectiveness and central velocity in fully developed region for meshes are listed in the Table 3.

Table 1: Thermo-physical properties of the (Cu-water) nanofluid

Material	**P Kg/m³**	**Cp J/Kg.K**	**K W/m.K**	μ **Kg/m.s**
Pure water c = 0%	981.3	4189	0.643	0.000598
clipper (Cu)	8930	383.1	386	

Cu-water c = 1%	1061	4150.9	0.662	0.000612
Cu-water c = 2%	1140.7	4112.8	0.682	0.000627
Cu-water c = 3%	1220.4	4074.8	0.702	0.000642
Cu-water c = 4%	1300.2	4036.7	0.723	0.000657
Cu-water c = 5%	1378	3998	0.744	0.000672

Table 2: Thermo-physical properties of the (Al_2O_3-water) nanofluid

Material	**P Kg/m³**	**Cp J/Kg.K**	**K W/m.K**	**μ Kg/m.s**
Pure water c = 0%	981.3	4189	0.643	0.000598
$Al^{2}O^{3}$	3600	765	36	-
$Al^{2}O^{3}$-water c = 1%	1007.4	4154.7	0.661	0.000612
$Al^{2}O^{3}$-water c = 2%	1033.6	4120.5	0.68	0.000627
$Al^{2}O^{3}$-water c = 3%	1059.8	4086.2	0.699	0.000642
$Al^{2}O^{3}$-water c = 4%	1086	4052	0.719	0.000657
$Al^{2}O^{3}$-water c = 5%	1112.2	4017.8	0.739	0.000672

Table 3 shows that the solution becomes independent of grid size and from second configuration further increase in the grids will not have a significant effect on the solution and results of such arrangement are acceptable. Therefore and to gate maximum accuracy the grid size of $(25 \times 25 \times 110)$ and $(25 \times 15 \times 110)$ is used.

The convergence criteria to control the solution for momentum and energy equation were set to be less than 10^{-6}.

Table 3: Grid independent study

Mesh Configuration	Effectiveness %	Fully Developed Central Velocity (m/s)
1.	42.81	0.6401
2	42.26	0.6489
3	42.22	0.6490

RESULTS AND DISCUSSION

First the model was solved with pure water as a cooling fluid with constant properties selected according to the mean temperature across channels [16, 17]. The inlet temperatures of hot and cold fluids used as boundary conditions are Thi = 373 K and Tci = 293 K.

To check the validity of the built numerical model, verification was made by solving the experimental model presented in [18] and results are compared. The experimental model presented in [18] is a microchannel heat sink consists of rectangular microchannels with hydraulic diameter 348.9 μm, channel height 713 μm, channel width 231 μm and length 4.48 cm. Temperature was measured in four points along the channel bottom wall, and experiment was made with inlet velocity of 1 m/s, inlet temperature of 288 K and thermal boundary condition is a constant heat flux of 100 W/cm^2 subjected on the bottom wall of the substrate.

Figure 3 shows the comparison between results of present numerical model and the experimental data of [18] for temperature distribution along the bottom wall of the channel. From this figure it can be seen that, the agreement between numerical and experimental results is acceptable since the maximum error is 1.41% which may be due to the end effect. Therefore the present numerical model is reliable and can be used to study the effect of using nanofluid as a coolant on the performance of a CFMCHE.

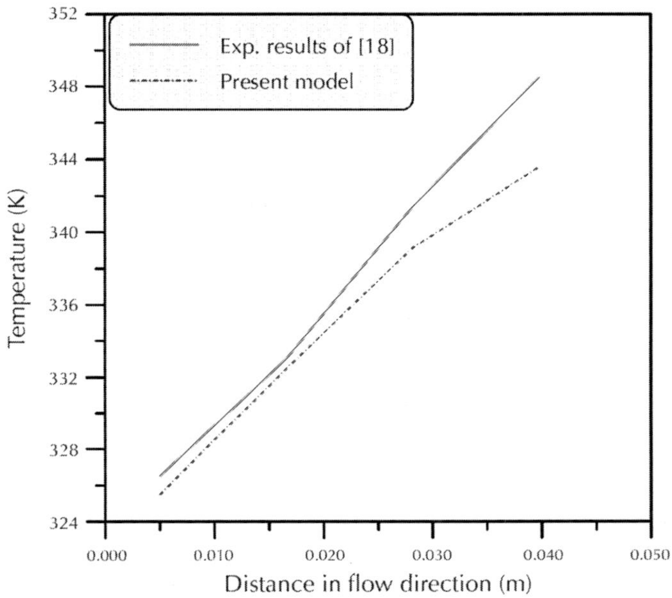

Figure 3: Variation of wall temperature in flow direction as a comparison between present model and [18].

Figure 4 shows the variation of thermal conductivity of nanofluid with particles volume fractions in the suspension. From this figure one can see that, thermal conductivity is increased with increasing volume fraction. Also it can be seen that the thermal conductivity of Cuwater nanofluid is higher than the conductivity of Al_2O_3- water nanofluid because the conductivity of copper is larger.

Variation of nanofluid viscosity with the volume fraction is indicated in Figure 5. From this figure the viscosity increased with increase of the fraction of solid particles in the suspension.

Figure 6 shows the variation of heat exchanger effectiveness with volume fraction for two types of nanofluids (Cu-water) and (Al_2O_3-water) at inlet velocity 1 m/s. From this figure it can be seen that, the effectiveness for the two nanofluids increased with increase the solid particles in the nanofluid. The increase of effectiveness with increase volume fraction is due to increase of thermal conductivity of nanofluids. Also the increasing in effectiveness in case of Cu-water is larger than that for Al_2O_3-water which is due to the higher thermal conductivity of Cu-water than Al_2O_3-water. The observed modification in effectiveness results from using of the two mentioned nanofluids compared with that for pure water is 9.86% for Cu-water and 5.04% for Al_2O_3-water at inlet velocity 1 m/s and volume fraction 5%.

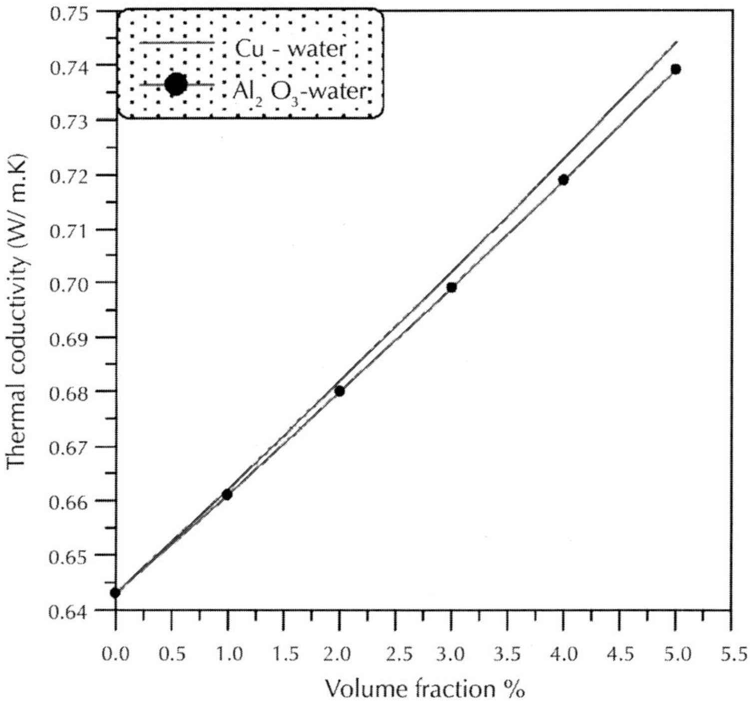

Figure 4: Variation of thermal conductivity with volume fraction for two types of nanofluids.

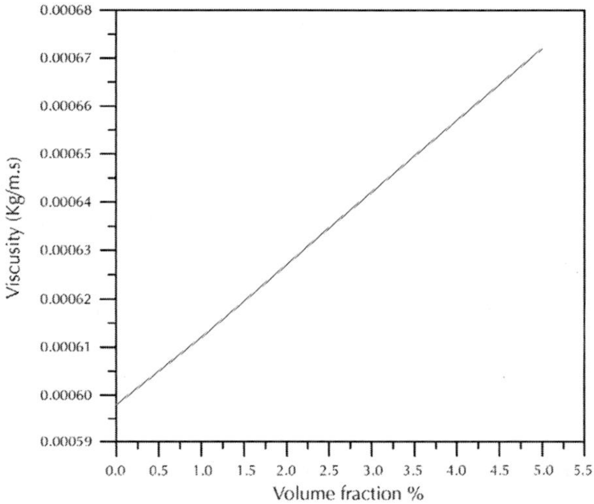

Figure 5: Variation of viscosity of nanofluid with volume fraction.

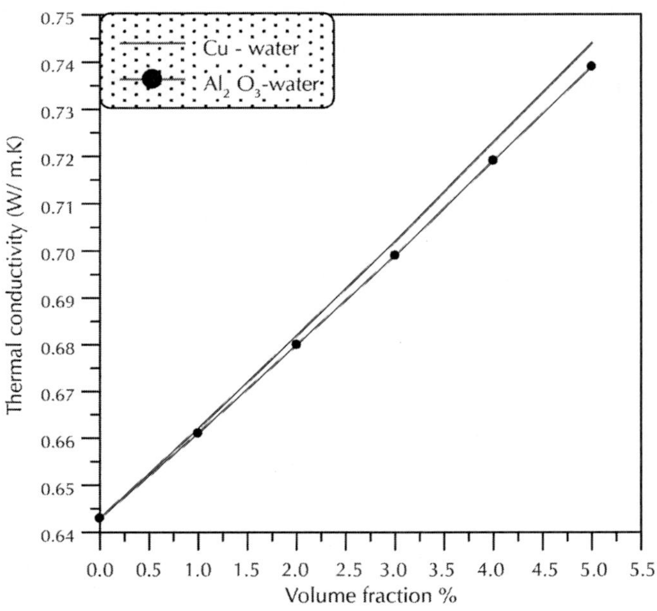

Figure 6: Variation of effectiveness with volume Fraction for two nanofluids at Vi = 1 m/s.

Figure 7 represents the variation of pressure drop across heat exchanger with volume fraction for the two nanofluids used. From this figure the pressure drop increased with increase volume fraction for nanofluids which is due to increase its viscosity. Also there is a small difference between pressure drop in Cu-water and Al_2O_3- water nanofluids which is due to the difference in density and viscosity between them.

The percentage increase in pressure drop associated with using nanofluids compared with pure water is 6.28% for Cu-water and 6.19% for Al_2O_3-water at volume fraction 5% for inlet velocity 1 m/s.

Variation of performance index ( ) with volume fraction for two nanofluids at Vi = 1 m/s is illustrated in Figure 8. From this figure one can see that, the performance index for two nanofluids increased with increase the volume fraction. For Cu-water nanofluid the increasing in performance index continue for all range of volume fraction chosen while for Al_2O_3-water the performance index reach a certain maximum value and then decreased. This is due to the increasing in effectiveness for Cu-water as a result of increase the thermal conductivity is larger than the increment in pressure drop due to increase in viscosity for all range of selected volume fractions. While in case of Al_2O_3-water this increment occurred for small range of volume fractions and for remaining range the effect of the increasing in pressure drop become larger than that of increasing in the effectiveness which leads to decrease the performance index.

The maximum percentage in enhancement of the performance index results from using nanofluids compared with pure water is 4.84% for Cu-water at volume fraction 5% and inlet velocity 1 m/s and 1.45% for Al_2O_3-water at volume fraction 1% and inlet velocity 1 m/s.

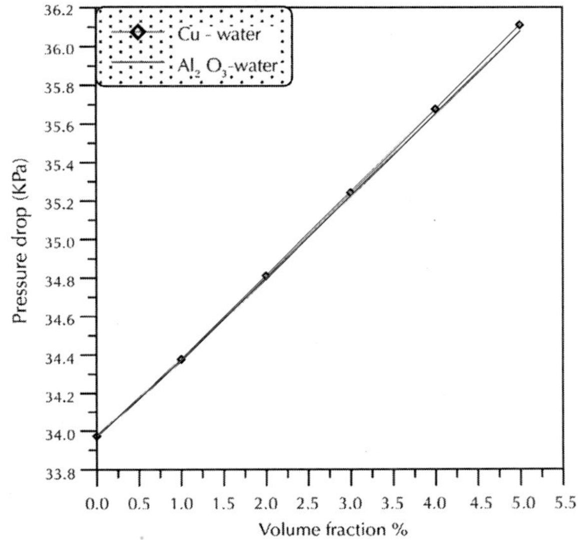

Figure 7: Variation of pressure drop with volume fraction for two nanofluids at Vi = 1 m/s.

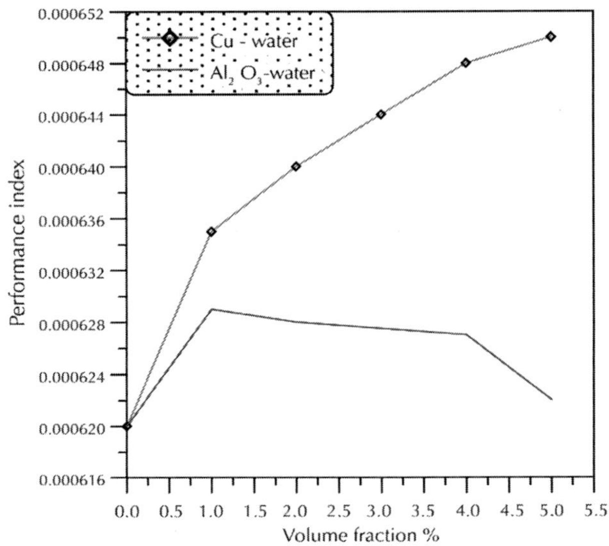

Figure 8: Variation of performance index with volume fraction for two nanofluids at Vi = 1 m/s.

Figure 9 illustrates the variation of heat transfer rate with the nanofluid volume fractions for two types of nanofluids. It's clear that the heat exchanged between two fluids increased with increasing of volume fraction. Also the heat transfer rate when using Cu-water is larger thanthat for Al_2O_3-water because Cu-water has higher thermal conductivity than Al_2O_3-water.

Variation of heat transfer rate over pumping power ratio (q/P.P) with volume fraction for two nanofluids at Vi = 1 m/s is shown in Figure 10. From this figure (q/P.P) ratio is increase with increase of volume fraction and has similar trend as for performance index shown in Figure **8**. The maximum modification in (q/P.P) compared with pure water is 4.57% for Cu-water at volume fraction 5% and Vi = 1 m/s and 1.23% for Al_2O_3-water at volume fraction 1% and Vi = 1 m/s.

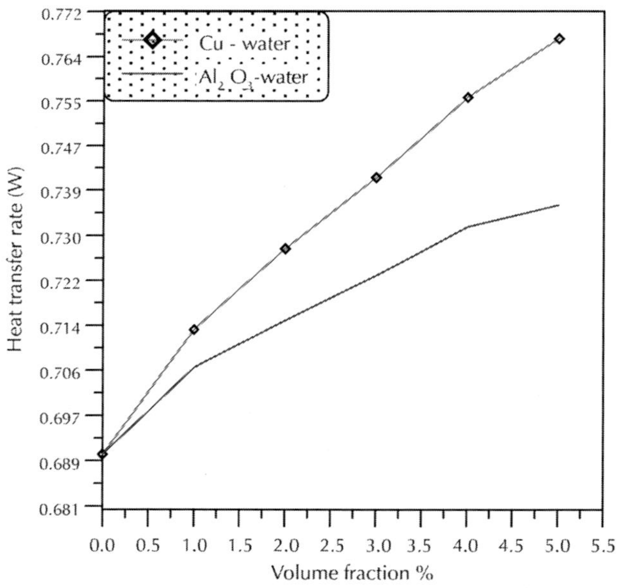

Figure 9: Variation of heat transfer rate with volume fraction for two nanofluids at Vi = 1 m/s.

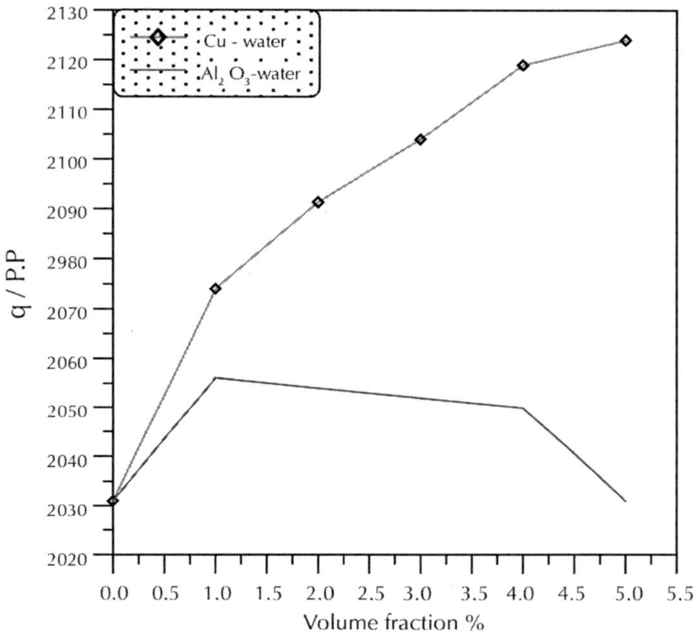

Figure 10: Variation of heat transfer rate over pumping power with volume fraction at Vi = 1 m/s.

Figure 11 shows the variation of effectiveness with inlet fluid velocity for different values of volume fraction. (c = 0 refer to pure water). From this figure it can be seen that the effectiveness decreased with increase of velocity. Also the large difference between different cases of volume fractions is occurred in low values of velocities and this difference is decreased with increase of fluid velocity. This is due to the effect of using nanofluids which appear in low flow rate while in high flow rate the flow is dominated by volume flow rate and the effect of sold particles on the developing of boundary layer.

Figure 12 shows the variation of performance index with inlet velocity for different values of volume fraction. From this figure the performance index decreased with increase inlet velocity. Also it can be seen that the difference between different cases is larger in low velocities and this difference decreases with increase of velocity due to decreasing of the effect of nanofluid in high velocity flow and the flow become dominated by flow rate. Figure 13 shows the variation of overall heat transfer coefficient along CFMCHE for different values of volume fractions. From this figure one can observes that the difference between different cases is larger in the nanofluid inlet side i.e. at the entrance region which means that existence of nanoparticles in a nanofluid affect the developing of boundary layer. Also it can be seen that the overall heat transfer coefficient is variable along heat exchanger due to the effect of entrance region and exist of axial conduction.

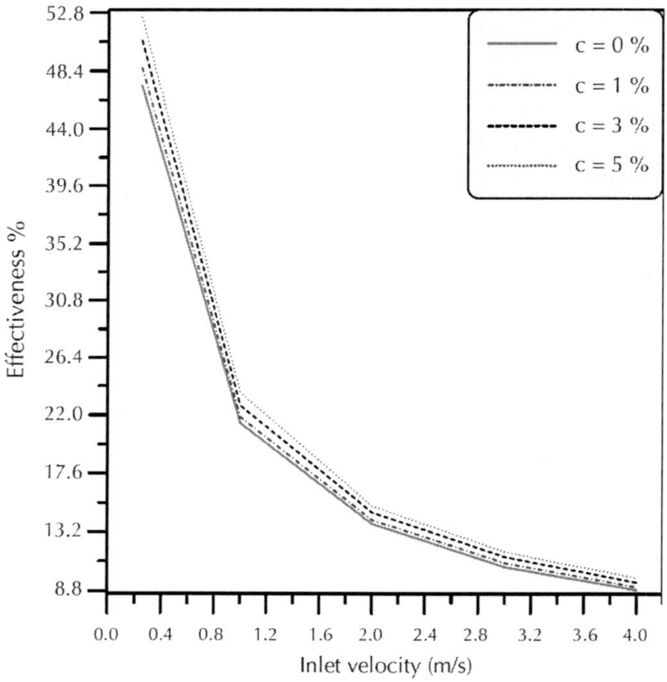

Figure 11: Variation of effectiveness with inlet velocity for different volume fractions for (Cu-water).

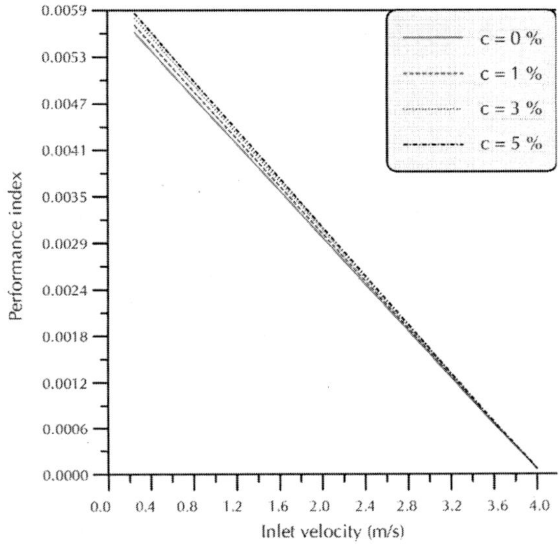

Figure 12: Variation of performance index with inlet velocity for different volume fractions for Cu-water.

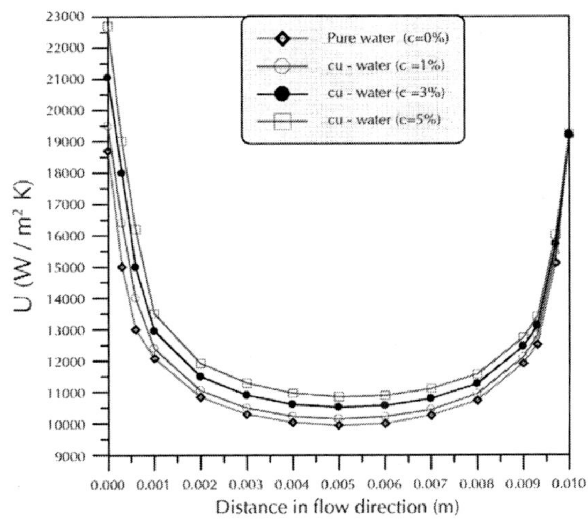

Figure 13: Variation of overall heat transfer coefficient along H.E for different volume fractions (Vi = 1 m/s).

CONCLUSIONS

In this paper a numerical investigation is made to study the effect of using nanofluids as cooling medium on the performance of counter flow microchannel heat exchanger. From the results it can be concluded that:

- Using of nanofluids lead to increase the effectiveness and cooling performance of a CFMCHE;
- There is no extra increase in pressure drop across CFMCHE associated with using of nanofluid as a coolant due to ultra-fine particles and small volume fractions;
- The effect of nanofluids is larger for low velocities while in flow with high velocities this effect is small because the flow will be dominated by the flow rate;
- The effect of nanofluids is high in the entrance region due to the effect of solid particles on the developing of boundary layer;
- The enhancement in thermal performance of CFM CHE increases with using material with high thermal conductivity.

REFERENCES

1. D. B. Tuckerman and R. F. W. Pease, "High-Performance Heat Sinking for VLSI," IEEE Electron Device Letters, Vol. 2, No. 5, 1981, pp. 126-129. doi:10.1109/EDL.1981.25367
2. X. W. Wang, X. F. Xu and S. U. S. Choi, "Thermal Conductivity of Nanoparticle-Fluid Mixture," Journal of Thermophysics and Heat Transfer, Vol. 13, No. 4, 1999, pp. 474-480.
3. S. Lee, S. U. S. Choi, S. Li and J. A. Estman, "Measuring Thermal Conductivity of Fluids Containing Oxide Nanoparticles," Journal of Heat Transfer, Vol. 121, No. 2, 1999, pp. 280-289. doi:10.1115/1.2825978
4. B.-X. Wang, L.-P. Zhou and X.-F. Peng, "A Fractal Model for Predicting the Effective Thermal Conductivity of Liquid with

Suspension of Nanoparticles," International Journal of Heat and Mass Transfer, Vol. 46, No. 14, 2003, pp. 2665-2672. doi:10.1016/S0017-9310(03)00016-4

5. J. Koo and C. Kleinstreuer, "A New Thermal Conductivity Model for Nanofluids," Journal of Nanoparticle Research, Vol. 6, No. 6, 2004, pp. 577-588. doi:10.1007/s11051-004-3170-5
6. Q. Li and Y. M. Xuan, "Convective Heat Transfer and Flow Characteristics of Cu-Water Nanofluid," Science in China Series E: Technological Sciences, Vol. 45, No. 4, 2002, pp. 408-416.
7. Y. M. Xuan and Q. Li, "Investigation on Convective Heat Transfer and Flow Features of Nanofluids," Journal of Heat Transfer, Vol. 125, No. 1, 2003, pp. 151-155.doi:10.1115/1.1532008
8. B. C. Pak and Y. I. Cho, "Hydrodynamic and Heat Transfer Study of Dispersed Fluids with Submicron Metallic Oxide Particles," Experimental Heat Transfer, Vol. 11, No. 2, 1998, pp. 151-170. doi:10.1080/08916159808946559
9. D. S. Wen and Y. L. Ding, "Experimental Investigation into Convective Heat Transfer of Nanofluids at the Entrance Region under Laminar Flow Conditions," International Journal of Heat and Mass Transfer, Vol. 47, No. 24, 2004, pp. 5181-5188.doi:10.1016/j.ijheatmasstransfer.2004.07.012
10. S. P. Jang and S. U. S. Choi, "Cooling Performance of a Microchannel Heat Sink with Nanofluids," Applied Thermal Engineering, Vol. 26, No. 17-18, 2006, pp. 2457-2463. doi:10.1016/j.applthermaleng.2006.02.036
11. T.-H. Tsai and R. Chein, "Performance Analysis of Nanofluid-Cooled Microchannel Heat Sinks," International Journal of Heat and Fluid Flow, Vol. 28, No. 5, 2007, pp. 1013-1026. doi:10.1016/j.ijheatfluidflow.2007.01.007
12. J. Lee and I. Mudawar, "Assessment of the Effectiveness of Nanofluids for Single-Phase and Two-Phase Heat Transfer in Micro-Channels," International Journal of Heat and Mass Transfer, Vol. 50, No. 3-4, 2007, pp. 452-463.doi:10.1016/j.

ijheatmasstransfer.2006.08.001

13. M. I. Hasan, A. A. Rageb, M. Yaghoubi and H. Homayony, "Influence of Channel Geometry on the Performance of Counter Flow Microchannel Heat Exchanger," International Journal of Thermal Sciences, Vol. 48, No. 8, 2009, pp. 1607-1618.doi:10.1016/j.ijthermalsci.2009.01.004
14. M. I. Hasan, "Numerical Investigation of Counter Flow Microchannel Heat Exchanger with MEPCM Suspension," Applied Thermal Engineering, Vol. 31, No. 6-7, 2011, pp. 1068-1075. doi:10.1016/j.applthermaleng.2010.11.032
15. R. Chein and J. Chuang, "Experimental Microchannel Heat Sink Performance Studies Using Nanofluids," International Journal of Thermal Sciences, Vol. 46, No. 1, 2007, pp. 57-66. doi:10.1016/j.ijthermalsci.2006.03.009
16. S. V. Patankar, "Numerical Heat Transfer and Fluid Flow," Hemisphere Publishing Corporation, Washington DC, New York, London, 1980.
17. F. P. Incropera and D. P. Dewitt, "Fundamentals of heat and mass transfer," 4th Edition, John Wiley and Sons, Hoboken, 1996.
18. Q. Weilin and M. Issam, "Thermal Design Methodology for High-Heat-Flux Single Phase and Two Phase Microchannel Heat Sinks," IEEE Transactions on Components and Packaging Technologies, Vol. 26, No. 3, 2003, pp. 598-609.doi:10.1109/TCAPT.2003.817652

Chapter 5

Experimental Analysis of Horizontal Ground Heat Exchanger for Northern Tunisia

Nabiha Naili, Issam Attar, Majdi Hazami, and Abdelhamid Farhat

Laboratory Thermal Processes, Research and Technology Centre of Energy, Hammam Lif, Tunisia

ABSTRACT

The aim of this study is first to evaluate the Tunisian geothermal energy and second to test the performance of horizontal ground heat exchanger. An experimental set-up has been constructed for climatic condition of Borj Cedria located in the north of Tunisia for space cooling. Results obtained during experiment were presented and discussed. The ground temperature at several depths was measured, the overall heat transfer coefficient (U) was determined. To evaluate the system efficiency, the energy analysis was applied;

the energy efficiency was found to range from 14% to 28%. The heat exchange rate was quantified, the pressure losses were calculated. The total heat rejected by using the ground heat exchanger (GHE) system was compared to the total cool requirements of a tested room with 12 m^2 surface. The results showed that the GHE, with 25 m of length buried at 1 m depth, covers 38% of the total cool requirement of the tested room. This study showed that the ground heat exchanger provide a new way of cooling buildings, it also showed that Tunisia have an important thermal potential. This favorable circumstance allows Tunisia to be a pioneer in the exploitation of geothermal energy for the installation of ground source heat pump systems.

INTRODUCTION

Tunisia, like other countries in the world, benefits from numerous natural geothermal sources. We can notice that this potential is not largely used and that, with the evolution of energy prices, geothermal energy as other renewable sources of energy must now considered with a new interest especially in the building heating and cooling.

Geothermal energy as environmentally friendly energy source with wide range of applications such as for space heating and cooling, hot water supply and applications in the agricultural field has been used in practical engineering. The well-known application is for space heating and cooling in residential and commercial buildings with using ground source heat pump system [1].

Ground-source heat exchangers provide a new and clean way of heating buildings in the world. They make use of renewable energy stored in the ground, providing one of the most energy-efficient ways of heating buildings. They are suitable for a wide variety of building types and are particularly appropriate for low environmental impact projects. They do not require hot rocks (geothermal energy) and can be installed in most of the world, using a borehole or shallow trenches or, less commonly, by extracting heat from a pond or lake. Heat collecting pipes in a closed loop, containing water (with a little antifreeze) are used to extract this stored energy, which can then be used to provide space heating and domestic hot water [2].

In the last decade the development of GHE technology has registered a great interest worldwide. The first working GSHP prototype was installed in the USA in 1945 as a demonstration device and it succeeded in showing clearly the big advantages of such an innovative solution. The ground heat exchanger (GHE) is an important part of the ground source heat pump (GSHP), it consists of a length of pipe buried at reasonable depth below the ground surface. There are many basic types of ground loop systems [3-6]. The GHE is best depends on the climate, soil conditions, available land, and local installation costs at the site.

In the open literature, many researches works have been conducted, modeling and testing of ground coupled heat pump systems [7-18].

Esen et al. [7] have tested the performance of an air-conditioning system formed by a ground coupled heat pump with two different depths for ground exchanger: 1 m and 2 m. Their experience showed that the ground exchanger performance increases with the depth (2.5 for 1 m and 2.8 for 2 m). They have also carried out a comparative economic evaluation [9]. Zhao et al. [10] indicate that heat transfer mainly happens near the outer wall of coaxial GHE and inclines to stabilization at far-field, they also have concluded that the inlet temperature, initial temperature of porous medium and the flow rate are major factors affecting heat transfer. Cui et al. [12] have developed a finite element numerical model to simulate hybrid ground-coupled heat pump with domestic hot water heating (DHW). Authors have concluded that The HGCHP can offer almost 95% of total DHW demand in this case study along with about 70% energy saving compared to the electric heater. Bi et al. [17] have employed a two-dimensional cylindrical coordinate system to model a ground heat exchanger. A CFD simulations have been carried out by Congedo et al. [18] to analyses three main geometries of horizontal ground heat exchangers: linear, helical and slinky. In their study, the authors have concluded that the most important parameter for the heat transfer performance of the system is the thermal conductivity of the ground around the heat exchanger, they have also showed that the choice of the velocity of the heat transfer fluid inside the tubes is a key factor for heat transfer performance for all the arrangements.

In this paper, an experimental set-up was constructed for climatic condition of Borj Cedria localized in the north of Tunisia. The aims of this study are first to evaluate the Tunisian geothermal energy and second to test the performance of horizontal ground heat exchanger with 25 m of length buried at 1 m depth.

EXPERIMENTAL SET-UP

Description of the System

The schematic arrangement of the experimental system is given in Figure 1. The experimental set-up consists mainly of two units: the ground heat exchanger (GHE), which its photograph is chow in Figure 2 and the heat pump. The GHE consists of a polyethylene tube, 16 mm in internal diameter, buried at 1 m depth. The distance between tubes is 0.3 m to minimize the interference between them. An air-liquid Clivet type heat pump was used to design tests. The specifications and characteristics of the components of the experimental system are given in Table 1.

Measurements Procedure

The experimental approach consists to fix the water temperature and mass flow rates at the inlet of the exchanger, when the temperature in the outlet side of the exchanger is stabilized (steady-state), we note the outlet temperature, which used to accurate the experimental analysis. Various tests were carried out in a range of mass flow rate between 0.06 and 0.2 $kg{\cdot}s^{-1}$.

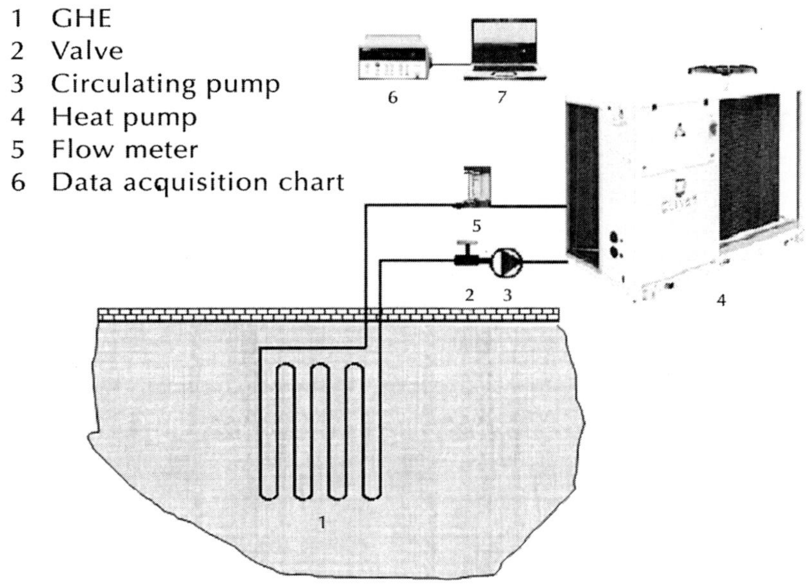

Figure 1: Schematic of the experimental system.

Figure 2: View of the GHE buried at 1 m depth.

Table 1: Experimental parameters

Heat pump information	
Capacity (kW)	21.9
Compressor type	Copeland Scroll,
Refrigerant type	R-22
Ground heat exchanger information	
Configuration type	Horizontal
Pipe materiel	Polyethylene
Length of pipe (m)	25
Internal diameter (m)	0.016
External diameter (m)	0.02
Piping depth (m)	1
Pipe distance (m)	0.3
Circulating pump information	
Type	Salamson
Power (W)	30

In the present study, the temperature and masse flow rates, were measured by appropriate instruments explained below:

- The inlet, outlet temperatures of GHE and the ground temperature were measured by using thermocouples.
- The water mass flow rate was measured by a rotameter.
- The pressures losses were determined with empirical formula in the literature [19].

All measured temperatures were recorded by using a multi-channel data-logger type HP, which were stocked in a PC. The experimental results were obtained from the ground-source heat pump-system in the cooling season of 2010-2011.

Uncertainty Analysis

Uncertainty analysis is needed to prove the accuracy of the experiments. In this study, errors came from the sensitiveness of equipment and measurements explained previously. First; errors due to measurement of temperature are: 1) sensitiveness of data acquisition system, about ± 0.1%˚C; 2) measurement error is ±0.2% and 3) sensitiveness of the thermocouple is ±0.1%˚C. The sensitiveness was obtained from a catalog of the instruments. Second, errors came from the measurement of the flow rate: 1) the sensitiveness of the flow meter is about ±0.1% and 2) errors due to measurement are about ±0.1%. In total, errors of measurement of the flow rate are about ±0.2%.

SOIL AND CLIMATE PROPERTIES

Soil Characteristics

In geothermal systems applications, deposition or extraction of thermal energy from the ground is accomplished by using a ground heat exchanger (GHE). The operation of this heat exchanger induces a simultaneous heat and moisture flow in the surrounding soil. The transfer of heat between the GHE and adjoining soil is primarily by heat conduction and to a certain degree by moisture migration. Therefore, it depends strongly on the soil type, temperature and moisture gradients [20].

The soil thermal diffusivity (a) is a defined property and is the ratio of the thermal conductivity (l) and the heat capacity (rC) (Table 2). Therefore, the three soil properties, l, r and C must be known or estimated to predict the thermal behavior of GHEs [21]. Obtaining accurate values for the thermal properties of the soil requires a detailed site survey. In order to estimate the thermal properties of granular soils (sand, clay, silt), it is necessary to determine the sand and clay content, dry density and moisture content [22].

Weather Data

The experimental set-up was established at the Research and Technology Center of Energy, Borj Cedria, northern Tunisia, latitude

36.83 and longitude 10.23. The climatic conditions of Borj Cedria for average values (monthly average minimum, maximum and mean outdoor temperature, the monthly averages of relative humidity, wind velocity, solar radiation and Average soil temperature at 1 m depth for the cooling season) are given in Table 3.

Table 2: Thermal properties and physics of the principal components of the ground [23]

	Density ρ **,kg·m⁻³**	**Heat capacity C, kJ·K⁻¹·kg⁻¹**	**Heat capacity** ρ **C, MJ·K⁻¹·m⁻³**	**Thermal conductivity** λ **, W·K⁻¹·m⁻¹**
Minerals (average)	$2.65 \cdot 10^3$	0.80	2.10	2.90
Organic substance	$1.30 \cdot 10^3$	1.90	2.47	0.25
Water	$1.00 \cdot 10^3$	4.20	4.20	0.585
Air	$1.25 \cdot 10^3$	1.00	1.25	0.023

ANALYSIS

In this study we will consider only the heat exchange which is making in stationary regime. The heat transfer inside a pipe is makes by forced convection. When we consider an infinitesimal element dx of a tube in the coolant flow direction, the heat exchange rate Q_e is given by the following expression:

$$\mathrm{d}Q_e = \dot{m}CpdTf\left(x\right) = U\left(x\right)\left(Tf\left(x\right) - Tg\right)\mathrm{d}S \tag{1}$$

By admitting that the overall heat transfer coefficient remains constant all along the exchanger, (U(x) = U), after integration, the preceding equation gives

$$\dot{m}Cp\mathrm{Ln}\left(\left(T_o - T_g\right)/\left(T_i - T_g\right)\right) = US \tag{2}$$

where T_i and T_o are the exchanger inlet and outlet temperature respectively.

The heat exchange rate Q_e was calculated from the following equation:

$$Q_e = \dot{m}Cp\Delta T$$

(1) and (2) give

$$Q_e = US\left(\Delta T / \left(\mathrm{Ln}\left(T_o - T_g\right) / \left(T_i - T_g\right)\right)\right) \quad (3)$$

Where represents a log mean temperature difference DTLM, so the heat exchange rate can be puts as follows

$$Q_e = USDTLM \quad (4)$$

To evaluate the performance of the GHE, we often use the energy efficiency concept. It is defined by the ratio of the really heat exchange rate (Q_e) and the theoretically possible of maximum heat exchange rate (Q_{Max}), it is expressed by:

$$\varepsilon = Q_e / Q_{(\max)} \quad (5)$$

with

$$Q_{\max} = \dot{m}Cp\Delta T' \quad (6)$$

The exchanger energy efficiency is written then:

$$\varepsilon = \Delta T / \Delta T' \quad (7)$$

The pressure losses must be calculated in order to be able to balance the various criteria the ones compared to the others. For the calculation of the pressure losses, we must calculate the linear and singular pressure losses

$$\Delta p_{\text{total}} = \Delta p_{\text{lin}} + \Delta p_{\text{sin}} \quad (8)$$

The linear pressure loss for a flow in a rectilinear control is determined in the following way [24]:

$$\Delta p = \Lambda \rho \mathrm{u}^2 L / 2D \quad (9)$$

The calculation of the loss ratio of load (L) depends on the nature of the flow, laminar or turbulent. This last gives place to more significant pressure losses. The singular pressure loss is defined by [24]:

$$\Delta p = \xi \rho u^2 / 2 \tag{10}$$

Table 3: The climatic conditions of Borj Cedria over cooling season of 2010-2011

	Month			
	May	**June**	**July**	**August**
Average outdoor temperature (°C)	20.3	23.8	27.3	27.2
Minimum outdoor temperature (°C)	9.5	11.6	16.5	17.2
Maximum outdoor temperature (°C)	31.3	36.9	39.8	38.9
Average relative humidity (%)	65.9	62.2	61.2	64.5
Average wind velocity (m/s)	3.6	3.7	3.3	3.1
Average solar radiation (KWH)	413.5	448.0	475.5	454.1
Average soil temperature at 1 m depth (°C)	19.8	23.2	24.5	25.2

RESULTS AND DISCUSSIONS

The ground temperature constitutes an essential data in the installation of GHEs. To determine the ground temperature, we installed in different ground levels, thermocouples which are connected to an acquisition system data. The ground temperature results at various depths (d) measured in summer (21-23 June 2010) is shown in Figure 3, we can note that the ground sees its temperature changes decreased exponentially with depth. This decrease diminished as the ground depth increases because the high thermal inertia of the ground. Measurements show that the ground temperature below a certain depth remains relatively constant. The temperature at 1m depth is about 23°C while the outdoor temperature is about 37°C. When the temperature at 1m depth was compared with the outdoor temperature, we established that Tunisia benefits from an important natural geothermal source. Figure 4 shows the curves of the heat exchange rate versus the log mean temperature difference (DTLM) for the following flows rate: 0.06, 0.08, 0.1, 0.12 and 0.16 $kg \cdot s^{-1}$ and for a variety of inlet temperature ranging between 30°C and 50°C.

The experimental variation of the heat exchange rate (Q_e) versus the log mean temperature difference (DTLM), is linear. The respective slopes of these lines brought back to the unit of surface are, in steady state, the overall heat transfer coefficient, U, of the exchanger.

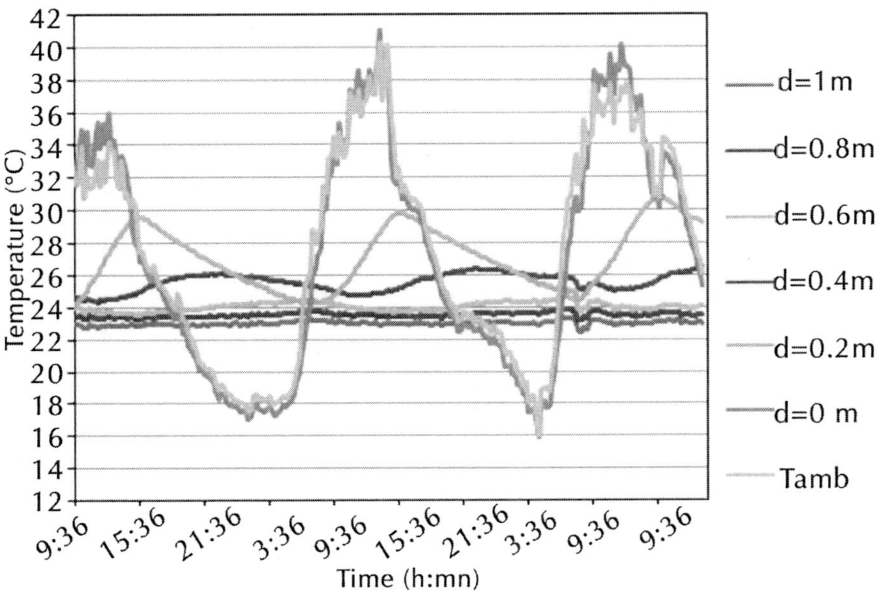

Figure 3: Soil temperature at several depths 21-23 June 2010.

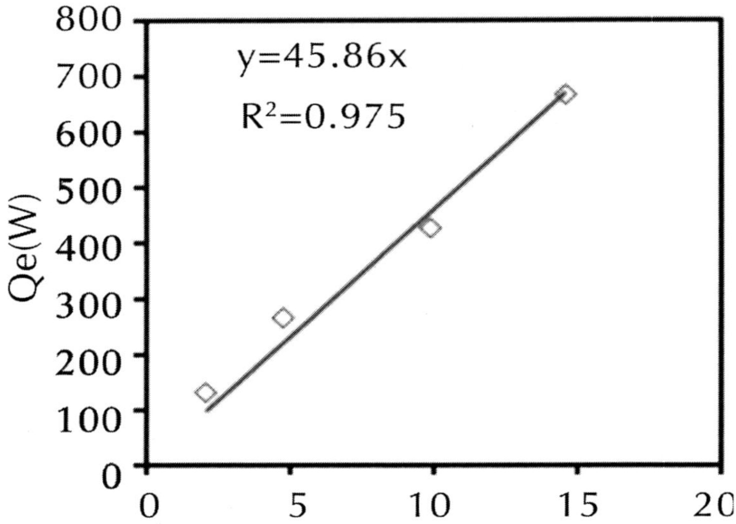
y=45.86x
R²=0.975
Qe(W)
800
700
600
500
400
300
200
100
0
0
5
10
15
20

(a)

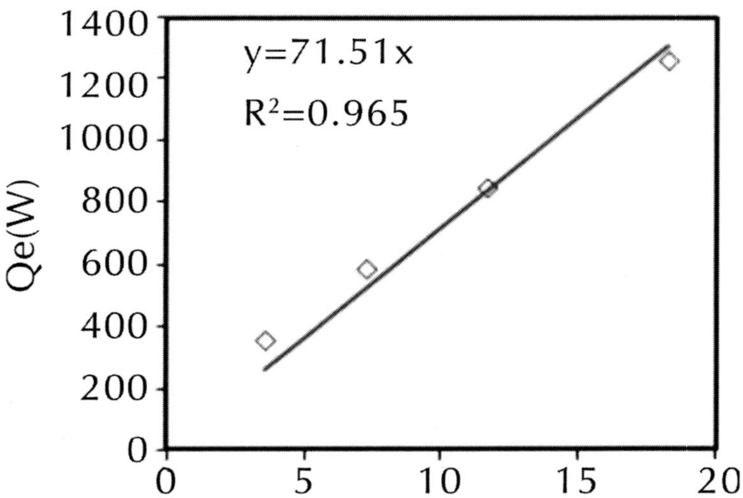
y=71.51x
R²=0.965
Qe(W)
1400
1200
1000
800
600
400
200
0
0
5
10
15
20

(b)

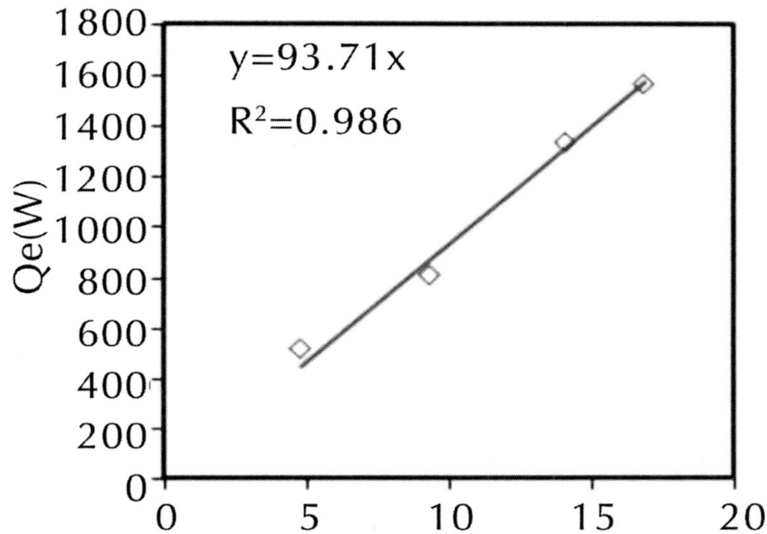
y=93.71x
R²=0.986
Qe(W)
1800
1600
1400
1200
1000
800
600
400
200
0
0
5
10
15
20

(c)

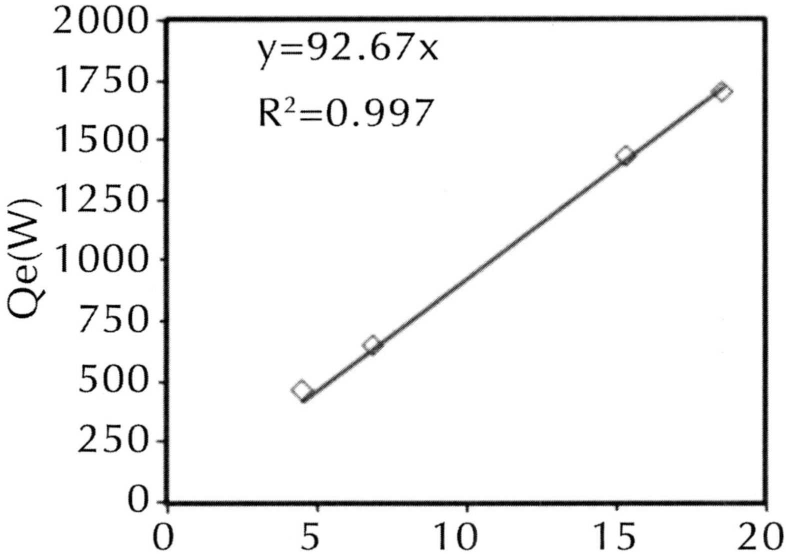
y=92.67x
R²=0.997
Qe(W)
2000
1750
1500
1250
1000
750
500
250
0
0
5
10
15
20

(d)

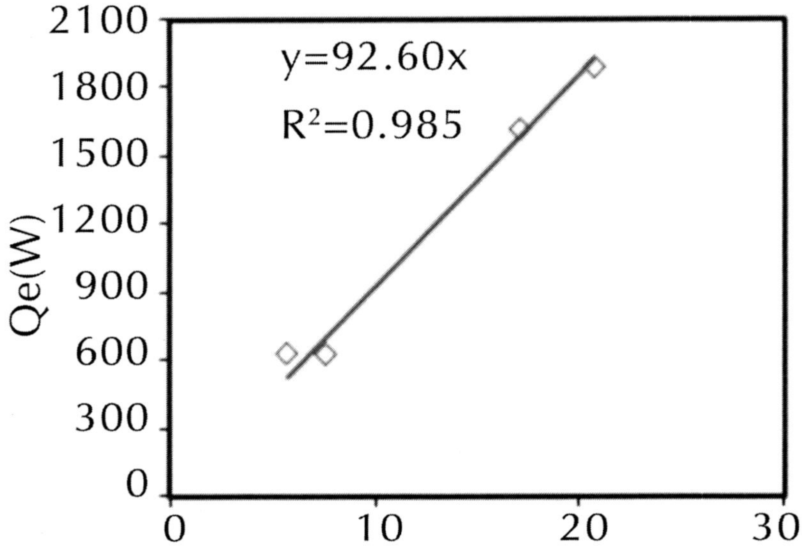

(e)

Figure 4: Heat exchange rate versus the log mean temperature difference for the following flow rates (a) 0.06 kg·s^{-1}; (b) 0.08 kg·s^{-1}; (c) 0.1 kg·s^{-1}; (d) 0.12 kg·s^{-1}; (e) 0.16 kg·s^{-1}.

The overall heat transfer coefficient values are regrouped in Table 4. We can conclude from Table 4 that the overall heat transfer coefficient, U, increases versus the mass flow rate. This increase is not linear. It is slowed down by the pressure losses. The overall heat exchange coefficient reached, during experiments, a maximum value of 59.02 W·m^{-2}·°C^{-1} for a 25 m of exchanger length and mass flow rate about 0.12 kg·s^{-1}. The use of the exchanger with lower flows decreases notably the overall heat transfer coefficient.

The use of the ground coupled heat pump aims essentially to have the maximum of the heat exchange rate, this quantity depends from many parameters. In our study we try to test the effect of the water coolant flow rate on this quantity, so we represent in Figure 5 the heat exchange rate Q_e variation versus mass flow rate.

It is noted that the heat exchange rate tends towards a limiting value about 650 W obtained for a water mass flow rate about 0.12 $kg{\cdot}s^{-1}$. Thus 0.12 $kg{\cdot}s^{-1}$ is the optimal mass flow rate. Indeed, any increase in the flow of water coolant beyond this value, does not affect the heat exchange rate. We can also concluded from this figure that the heat exchange rate, when the temperature in the outlet side of the exchanger is stabilized (steady-state) and for the masse flow rate about 0.12 $kg{\cdot}s^{-1}$, is about 26 W/m witch reflects the importance of geothermal energy in Tunisia.

The energy efficiency of the ground heat exchanger is determined by the NTU, the results are shown in Figure 6. It is noticed from this figure that the energy efficiency increases with NTU, which ranging from 14.3% to 28.4%. Indeed, the NTU determines the residence time of the water on the heat exchanger surface and thus also the amount of heat that it can absorbs by the ground. Consequently when the NTU increases the residence time of the water in the heat exchanger increases, thus the outlet temperature increases witch increase the energy efficiency.

To calculate the pressure loss we must first calculate the Reynolds number R_e (Table 5) to estimate the type of flow for the various flows used.

It is seen, in the Table 5, that for the range of the flows which we considered in our experimental study, it is difficult to obtain a laminar flow. We will thus consider for the continuation the pressure loss for a turbulent flow. In this case, the loss ratio of load can be determined by the Blasius relation:

In Figure 7 we represent the variation of the pressure loss versus mass flow rate. It is noticed that more the mass flow rate increases more the pressure loss increases too (Equations (9) and (10)). This is due to the several accidents met by the liquid coolant inside the exchanger. For the experimental masse flow rate (0.12 $kg{\cdot}s^{-1}$) the pressure loss is about 8.43 kPa.

This result is compatible with the experimental and numerical study published in the literature (Missirlis et al. [19]).

Table 4: The overall heat transfer coefficient

Mass flow rate ($kg \cdot s^{-1}$)	0.06	0.08	0.1	0.12	0.16
U($W \cdot m^{-2}$ °C)	29.15	45.04	58.4	59.02	58.98

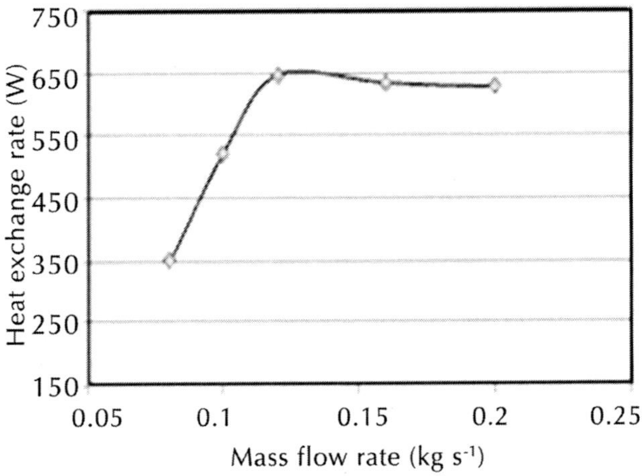

Figure 5: Heat exchange rate versus the flow rate for Ti = 30°C.

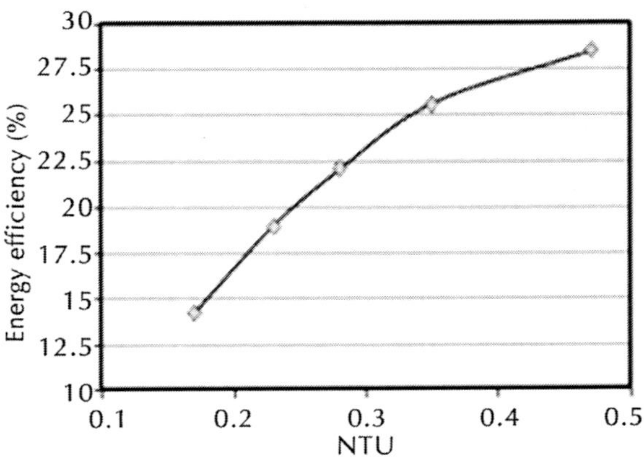

Figure 6: Energy efficiency versus NTU.

Table 5: Calculation of the Reynolds number

Mass flow rate (Kg s^{-1})	**u(m.s^{-1})**	**R_e**
0.06	0.299	5941
0.08	0.39	7922
0.1	0.49	9902
0.12	0.59	11,883
0.15	0.74	14,854

The total heat rejected by using the HGSHE system was compared to the total cool requirements of a tested room (4 m × 3 m × 3 m), which the temperature of thermal comfort under cover is 26°C in summer, in climatic condition of cooling season of Tunisia. We established that for the optimal flow rate procurement found (0.12 kg·s^{-1}), the maximum thermal power extracted from the ground is about 650 W. this value represents 38.5% of the cooling loads of the room test. The results showed that Tunisia have an important thermal potential. This favourable circumstance allows Tunisia to be a pioneer in the exploitation of geothermal energy for the installation of ground source heat pump systems.

CONCLUSIONS

In the present study horizontal ground heat exchanger was buried at 1 m and tested in the Research and Technology Center of Energy, Tunisia. The following conclusions can be drawn from this study:

- Thermal potential in Tunisia offers a good exploitation of horizontal ground source heat exchanger (HGSHE), indeed the experimental ground temperature shows that, at sufficient depth, it is always lower than that of the outside air in summer. It can be seen also that the ground temperature is nearly constant below a depth of 1 m;
- The energy efficiency for the GSHE system considered are found to range from 14.3% to 28.4%;

- The heat exchange rate, when the temperature in the outlet side of the exchanger is stabilized (steady-state), is about 26 W/m witch reflects the importance of surface geothermal energy in Tunisia;
- The pressure loss increase with the mass flow rate, for the optimal mass flow rate obtained (0.12 $kg \cdot s^{-1}$) the pressure loss is about 8.43 kPa;
- The GHE, with 25 m of length buried at 1m depth, covers 38.5% of the total cool requirement of a tested room with 12 m^2 surface.

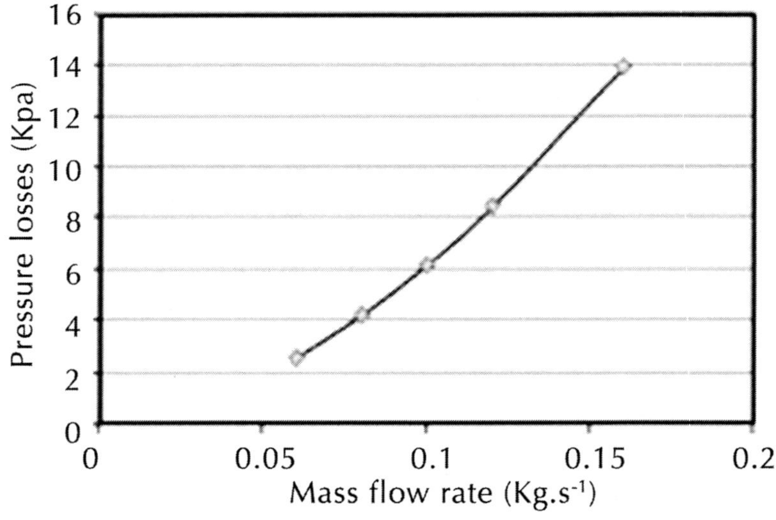

Figure 7: Pressure loss versus mass flow rate.

ACKNOWLEDGMENTS

The authors would like to thank the Laboratoire des Procédés Thermiques (LPT) for financially supporting the project and for supplying some useful data.

REFERENCES

1. Jalaluddin and A. Miyara, "Thermal Performance Investigation of Several Types of Vertical Ground Heat Exchangers with Different Operation Mode," Applied Thermal Engineering, Vol. 33-34, 2012, pp. 167-174. doi:10.1016/j.applthermaleng.2011.09.030
2. A. M. Omer, "Ground-Source Heat Pumps Systems and Applications," Renewable and Sustainable Energy Reviews, Vol. 12, No. 2, 2008, pp. 344-371.doi:10.1016/j.rser.2006.10.003
3. G. Axelsson, "Sustainable Geothermal Utilization—Case Histories; Definitions; Research Issues and Modelling," Geothermics, Vol. 39, No. 4, 2010, pp. 283-291.doi:10.1016/j.geothermics.2010.08.001
4. X. H. Xu, S. W. Wang, J. B. Wang and F. Xiao, "Active Pipe-Embedded Structures in Buildings for Utilizing LowGrade Energy Sources: A Review," Energy and Buildings, Vol. 42, No. 10, 2010, pp. 1567-1581. doi:10.1016/j.enbuild.2010.05.002
5. G. Florides and S. Kalogirou, "Ground Heat Exchangers—A Review of Systems, Models and Applications," Renewable Energy, Vol. 32, No. 15, 2007, pp. 2461-2478.doi:10.1016/j.renene.2006.12.014
6. D. Mileni , P. Vasiljevi and A. Vranješ, "Criteria for Use of Groundwater as Renewable Energy Source in Geothermal Heat Pump Systems for Building Heating/ Cooling Purposes," Energy and Buildings, Vol. 42, No. 5, 2010, pp. 649-657. doi:10.1016/j.enbuild.2009.11.002
7. H. Esen, M. Inalli, M. Esen and K. Pihtili, "Energy and Exergy Analysis of a Ground-Coupled Heat Pump System with Two Horizontal Ground Heat Exchangers," Building and Environment, Vol. 42, No. 10, 2007, pp. 3606-3615. doi:10.1016/j.buildenv.2006.10.014
8. U. Eicker and C. Vorschulze, "Potential of Geothermal Heat Exchangers for Office Building Climatisation," Renewable

Energy, Vol. 34, No. 4, 2009, pp. 1126-1133.doi:10.1016/j.renene.2008.06.019

9. H. Esen, M. Inalli and M. Esen, "A Techno-Economic Comparison of Ground-Coupled and Air-Coupled Heat Pump System for Space Cooling," Building and Environment, Vol. 42, No. 5, 2007, pp. 1955-1965. doi:10.1016/j.buildenv.2006.04.007

10. J. Zhao, H. J. Wang, X. G. Li and C. S. Dai, "Experimental Investigation and Theoretical Model of Heat Transfer of Saturated Soil around Coaxial Ground Coupled Heat Exchanger," Applied Thermal Engineering, Vol. 28, No. 2-3, 2008, pp. 116-125.

11. C. A. De Swardt and J. P. Meyer, "A Performance Comparison between an Air-Source and a Ground-Source Reversible Heat Pump," International Journal of Energy Research, Vol. 25, No. 10, 2001, pp. 899-910. doi:10.1002/er.730

12. P. Cui, H. Yang and Z. Fang, "Numerical Analysis and Experimental Validation of Heat Transfer in Ground Heat Exchangers in Alternative Operation Modes," Energy and Buildings, Vol. 40, No. 6, 2008, pp. 1060-1066. doi:10.1016/j.enbuild.2007.10.005

13. H. Demir, A. Koyun and G. Temir, "Heat Transfer of Horizontal Parallel Pipe Ground Heat Exchanger and Experimental Verification," Applied Thermal Engineering, Vol. 29, No. 2-3, 2009, pp. 224-233. doi:10.1016/j.applthermaleng.2008.02.027

14. A. Negiz, M. A. Hastaoglu and R. A. Heidemann, "ThreeDimensional Transient Heat Transfer from a Buried Pipe —I. Laminar Flow," Chemical Engineering Science, Vol. 48, No. 20, 1993, pp. 3507-3517. doi:10.1016/0009-2509(93)85006-B

15. A. Negiz, M. A. Hastaoglu and R. A. Heidemann, "ThreeDimensional Transient Heat Transfer from a Buried Pipe: Solidification of a Stationary Fluid," Numerical Heat Transfer, Vol. 28, No. 2, 1995, pp. 175-193.

doi:10.1080/10407789508913740

16. S. Thiers and B. Peuportier, "Modélisation Thermique d'un échangeur Air/Sol Pour le Rafraîchissement des Bâtiments," Journée Thématique SFT-IBPSA Froid Solaire et Confort d'été, 2007, Aix-les-Bains.
17. Y. H. Bi, L. G. Chen and C. Wu, "Ground Heat Exchanger Temperature Distribution Analysis and Experimental Verification," Applied Thermal Engineering, Vol. 22, No. 2, 2002, pp. 183-189. doi:10.1016/S1359-4311(01)00073-4
18. P. M. Congedo, G. Colangelo and G. Starace, "CFD Simulations of Horizontal Ground Heat Exchangers: A Comparison among Different Configurations," Applied Thermal Engineering, Vol. 33-34, 2012, pp. 24-32. doi:10.1016/j.applthermaleng.2011.09.005
19. D. Missirlis, S. Donnerhack, O. Seite, C. Albanakis, A. Sideridis, K. Yakinthos and A. Goulas, "Numerical Development of a Heat Transfer and Pressure Drop Porosity Model for a Heat Exchanger for Aero Engine Applications," Applied Thermal Engineering, Vol. 30, No. 11-12, 2010, pp. 1341-1350. doi:10.1016/j.applthermaleng.2010.02.021
20. W. H. Leong, V. R. Tarnawski and A. Aittomaki, "Effect of Soil Type and Moisture Content on Ground Heat Pump Performance," International Journal of Refrigeration, Vol. 21, No. 8, 1998, pp. 595-606. doi:10.1016/S0140-7007(98)00041-3
21. J. E. Bose, J. D. Parker and F. C. McQuiston, "Design/ Data Manual for Closed-Loop Ground-Coupled Heat Pump Systems," American Society of Heating, Refrigerating and Air-Conditioning Engineers, Atlanta, 1985.
22. S. P. Kavanaugh and K. Rafferty, "Ground-Source Heat Pumps: Design of Geothermal Systems for Commercial and Institutional Buildings," American Society of Heating, Refrigerating and Air-Conditioning Engineers, Chiba, 1997.
23. M. Hazami, "Étude Expérimentale et Numérique du Phénomène de Stockage et de Déstockage de L'énergie Sensible

en Utilisant un Echangeur Capillaire en Polypropylène," Thèse de Doctorat, Faculté des Sciences de Tunis, Tunis, 2008.

24. I. E. IdelCik, "Mémento des Pertes de Charges: Coefficients de Pertes de Charges Singulières et de Pertes de Charge par Frottement," Traduit par M. Meury, Editions Eyrolles, Saint-Germain, Paris, 1986.

Chapter 6

Delaware Method Improvement for the Shell and Tubes Heat Exchanger Design

Miguel Toledo-Velázquez[1], Pedro Quinto-Diez[1], Juan C. Alzelmetti-Zaragoza[2], Sergio R. Galvan[3], Juan Abugaber-Francis[1], and Arturo Reyes-León[1]

[1]Applied Thermal and Hydraulic Engineering Laboratory SEPI-ESIME-IPN Professional Unit "Adolfo Lopez Mateos", México DF, México

[2]Faculty of Mechanical Electrical Engineering, Universidad Veracruzana, Veracruz, México

[3]Faculty of Mechanical Engineering, Universidad Michoacana de Sán Nicolas de Hidalgo, Edifice H, University City, Santiago Tapia 403, Col. Centro, Morelia, Michoacán, México

ABSTRACT

In this paper the Delaware Method published in 1963 is analized and upgraded with using correction factors which take into account the undesirable currents of the mean flow. However, this method presents graphically these correction factors which imply an impediment to fulfill the software calculations. Thus, the equations corresponding to the correction factor equations and a Fortran 77 numerical program were established. This system is given to explore different design alternatives in order to find the optimal solution to each proposed problem. The results of this work was a simple software that can perform calculations with the introduction of parameters depending only on the geometry of the heat exchanger, i.e., geometry, temperature and fluid characteristics eliminating the human errors and increasing the calculations speed and accuracy.

INTRODUCTION

The method is established in the heat transfer analysis and in the pressure losses of the fluid which flows through the shell side [1] [2] .

In order to complete the heat exchanger analysis, it is essential to consider the different currents generated by the shell side flow, as is shown in Figure 1. This figure has been modified by Palen and Taborek [3] [4] in regard to the original version proposed by Tinker [5] . Here five different streams are identified side of the shell. The current B is known as the mean mixed flow and flows through the mixed flow section and the window section of the heat exchanger. This is the ideal current flowing in the shell side of a heat exchanger.

Besides of flow B, there are four more flows which are present due to the free spaces and cavities between the shell and the deflectors and they provoke a modification of the current B performance.

The different leakages and recirculation currents influence the heat flux transfer in two ways:

- Reduce the current B resulting in a drop of the heat transfer global coefficient.
- Modify the shell-side temperature distribution.

The Delaware method considers these effects as correction factors in the heat transfer coefficient and the loss pressure calculations.

The basic equations of the heat exchangers thermal design are:

$$Q = U_{dc} A F T_{mlc} \tag{1}$$

$$U_{dc} = \frac{1}{\frac{1}{h_{io}} + \frac{1}{h_{cc}} + R_d + \frac{e_t}{k}} \tag{2}$$

In the Equation (2), h_{cc} is the fluid convection coefficient flowing in the shell and it is obtained by Equation (3). In this equation the correction factors are included, the correction factors J_i, J_c, J_l, J_b, J_r, J_r^* and J_s, which consider the different currents shown in Figure 1.

$$h_{cc} = J_i Cp_c \left(\frac{W_c}{S_m}\right)\left(\frac{k_c}{Cp_c \mu_c}\right)^{2/3}\left(\frac{\mu_c}{\mu_{wc}}\right)^{0.14} J_c J_l J_b J_r J_s \tag{3}$$

For the hydraulic design is required to determine the fluid pressure loss through the shell side and it must be included the correction factors f_i, R_b, R_l and R_s. The corresponding calculation is made using the next equation:

$$\Delta P_c = \left\{(N_b - 1)\left[\frac{4 f_i W_c^2 N_c}{2 g \rho_c S_m^2}\left(\frac{\mu_{wc}}{\mu_c}\right)^{0.14}\right] R_b R_l + 2\Delta P_{b,i}\left(1 + \frac{N_{cw}}{N_c}\right) R_b R_s + N_b \Delta P_{w,i} R_l\right\} \tag{4}$$

It is observed that the principal problem using the Delaware Method is the determination of the eleven correction factors of which nine are obtained graphically. In the next section, the nine correction factors will be presented analytically. The original version only presents analytically the J_s and R_s correction factors.

SHELL SIDE CORRECTION FACTOR EQUATIONS

According to the Delaware method, after the geometrical parameters computations, the heat transfer and loss pressure are estimated considering the correction factors. This special case will be shown in the next section by using the appropriate Equations [6] -[8] .

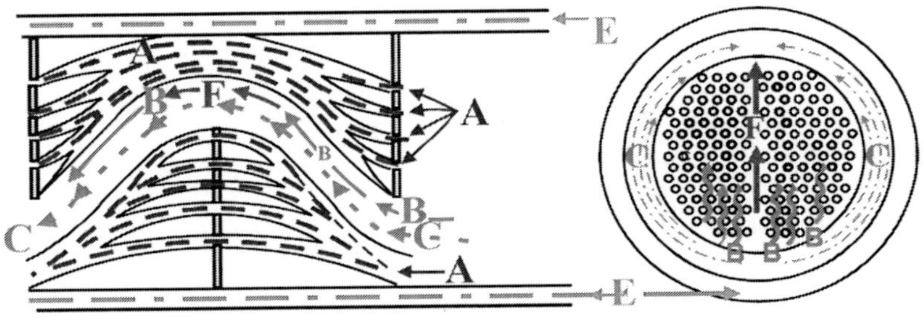

Figure 1: Ideal design of the currents flowing in the shell side. A. Currents flowing through the free spaces between tubes and baffles; B. Mean current flow; C. Recirculation current; D. Leakage current between shells and deflectors; E. Recirculation current in the past partition.

Correction Factor J_i

This factor depends on the Reynolds Number Re_c and the tube arrangement as is established in the next equation:

$$J_i = \exp\left[A + B\ln(Re_c) + C\ln(Re_c)^2 + D\ln(Re_c)^3 + E\ln(Re_c)^4 + F\ln(Re_c)^5\right]a \quad (5)$$

This equation is applied to the different arrangement tube types: triangular, square, and rhombic. For each case, the constant are presented in Table 1. It was determined 4% as that the maximum error between the equations and the graphic presentation.

Correction Factor J_c by the Baffle Configuration Effect

The deflector geometry and especially the window cross-section shape provoke currents where the effects are considered in the correction factor J_c. This factor is determined in a mathematical way as following:

$$J_c = A + BF_c + CF_c^2 + DF_c^3 + EF_c^4 \tag{6}$$

The values of each constant are presented in Table 2. Comparing the graphics obtained by Equation (6) and the original one, the maximum error established is 5%.

Correction Factor by Baffle Leakages Effect, J_l

The fluid that does not exchange heat owing to the leakage through the gap formed by the tubes and the baffles as well as through the shell and baffle is considered in the correction factor J_l. This factor is obtained using the Equation (7) where:

$$S_1 = S_{sb}/(S_{sb} + S_{tb}) \text{ and } S_2 = (S_{sb} + S_{tb})/S_m$$

The constants required by this equation are available in Table 3. Comparing the original graph against that one getting by Equation (7), the maximum error is 2%.

$$\begin{aligned} J_l = & \left[A + B(S_1) + C(S_1)^2 + D(S_1)^3\right] + \left[E + F(S_1) + G(S_1)^2 + H(S_1)^3\right](S_2) \\ & + \left[I + J(S_1) + K(S_1)^2 + L(S_1)^3\right](S_2)^2 + \left[M + N(S_1) + O(S_1)^2 + P(S_1)^3\right](S_2)^3 \end{aligned} \tag{7}$$

Table 1: Constants for the equation for the correction factor J_i

Arrangement	A	B	C	D	E	F
Triangular	0.627615	−0.69064	−0.0507472	0.0141049	−0.000937714	1.7683×10^{-5}
Square	0.374177	−0.671577	−0.0784051	0.02507191	−0.00224983	0.0000673254
Rhombic	−0.273166	−0.472896	−0.109701	0.023299	−0.00145983	0.0000242675

Table 2: Constants for the equation for the correction factor J_c

For values of F_c between	A	B	C	D	E
0 and 0.9	0.533574545	0.69059596	0.290909091	−0.295959596	0
0.9 and 1	−27.84837787	152.5274893	−301.9699773	265.12743360	−86.76640715

Table 3: Constants for the equation for the correction factor J_l

Values of S_2	A	B	C	D	E	F	G	H
0 to 0.1	1	0	0	0	−2.5903333	−4.8677761	8.773331	−7.0222218
0.1 to 0.7	0.90063003	−0.0389402	0.1363475	0.05531749	−0.4299145	−1.0949101	1.4689544	0.7341744
Values of S_2	**I**	**J**	**K**	**L**	**M**	**N**	**O**	**P**
0 to 0.1	15.8742857	50.8190476	−133.48571	106.666666	−39.6381	−168.31699	486.093325	−391.10933
0.1 to 0.7	−0.0567309	1.72685112	−4.2124254	2.4290305	0.04115536	1.3220234	3.32880672	−1.99187

Correction Factor by Recirculation Flow Effect, J_b

The recirculation currents do not exchange heat through the tubes and this issue is considered by the correction factor J_b. This factor is computed by Equation (8) which is based on F_{sbp}. The constant of the Equation $N_2 = N_{ss}/N_c$ are defined in Table 4 in terms of Re_c.

$$J_b = \left[A + B(N_2) + C(N_2)^2 + D(N_2)^3 + E(N_2)^4\right] + \left[F + G(N_2) + H(N_2)^2 + I(N_2)^3 + J(N_2)^4\right] F_{sbp} + \left[K + L(N_2) + M(N_2)^2 + N(N_2)^3 + \tilde{N}(N_2)^4\right] F_{sbp}^2 + \left[O + P(N_2) + Q(N_2)^2 + R(N_2)^3 + S(N_2)^4\right] F_{sbp}^3 + \left[T + U(N_2) + V(N_2)^2 + W(N_2)^3 + X(N_2)^4\right] F_{sbp}^4 \quad (8)$$

A maximum error of 1.3% is result of comparing the original graph against that one obtained by Equation (8).

Correction Factor by Adverse Temperature Gradient J_r

This factor has the value of 1 if Re_c ³ 100. When Re_c fluctuation is between 0 and 100, the next criterion is used:

- If Re_c £ 20, J_r will acquire the same value as J_r^* using the Equation (9), and knowing that $N_b$and $N_1 = N_c + N_{cw}$ are constants which are shown in Table 5.

$$J_r^* = \left[A + B(N_1) + C(N_1)^2 + D(N_1)^3 + E(N_1)^4\right] + \left[F + G(N_1) + H(N_1)^2 + I(N_1)^3 + J(N_1)^4\right] N_b + \left[K + L(N_1) + M(N_1)^2 + N(N_1)^3 + \tilde{N}(N_1)^4\right] N_b^2 + \left[O + P(N_1) + Q(N_1)^2 + R(N_1)^3 + S(N_1)^4\right] N_b^3 + \left[T + U(N_1) + V(N_1)^2 + W(N_1)^3 + X(N_1)^4\right] N_b^4 \quad (9)$$

A maximum error of 1.9% is result of comparing the original graph against that one obtained by Equation (8).

- If 20 £ Re_c £ 100, J_r is computed by Equation (10) which depends on J_r^*, Re_c and Table 6constants.

$$J_r = \left(A + B\,Re_c + C\,Re_c^2 + D\,Re_c^3\right) + \left(E + F\,Re_c + G\,Re_c^2 + H\,Re_c^3\right) J_r^* + \left(I + J\,Re_c + K\,Re_c^2 + L\,Re_c^3\right) J_r^{*2} + \left(M + N\,Re_c + O\,Re_c^2 + P\,Re_c^3\right) J_r^{*3} \quad (10)$$

In the previous equation, J_r^* is calculated by Equation (9). Comparing both graphics, the original and that one obtained by

Equation (10), it is observed as maximum error 3.8%.

Table 4: Constants for the equation for the correction factor J_b

For	**A**	**B**	**C**	**D**	**E**	**F**	**G**	**H**	**I**
$Re_c > 100$	0.99939474	0.01085613	−0.0171267	−0.1003071	0.19198207	−1.2394301	13.4729632	−85.593388	252.064457
$Re_c < 100$	0.9991889	0.02071706	−0.185339	0.59250929	−0.5964414	−1.3564474	16.400326	−116.51387	362.367998
For	**J**	**K**	**L**	**M**	**N**	**Ñ**	**O**	**P**	**Q**
$Re_c > 100$	−249.7121	0.74529366	−9.5329177	82.1849019	−321.67178	378.959635	−0.3575412	3.12005087	−71.95138
$Re_c < 100$	−368.18466	1.18264455	−19.614535	140.62959	−451.59784	478.689065	−1.1031648	16.9758369	−113.6298
For	**R**	**S**	**T**	**U**	**V**	**W**	**X**		
$Re_c > 100$	422.592705	−576.65103	0.20001675	−3.866875	70.1962129	−363.1866875	473.343386		
$Re_c < 100$	396.869404	−457.40996	0.59140536	−8.7366322	56.8613935	−210.10305	253.214998		

Table 5: Constants for the equation for the correction factor J_r^*

A	**B**	**C**	**D**	**E**	**F**	**G**	**H**	**I**
1.11286866	−0.0444079	0.0019872	-4.324×10^{-5}	3.4787×10^{-7}	−0.0289065	2.241×10^{-4}	1.5142×10^{-6}	9.5563×10^{-8}
J	**K**	**L**	**M**	**N**	**Ñ**	**O**	**P**	**Q**
-3.11×10^{-9}	0.0008806	2.8301×10^{-5}	-2.088×10^{-6}	3.4527×10^{-8}	-9.296×10^{-11}	-1.363×10^{-5}	-9.749×10^{-7}	5.9493×10^{-8}

R	S	T	U	V	W	X		
-8.947×10^{-10}	1.618×10^{-12}	8.3072×10^{-8}	8.7699×10^{-9}	-4.727×10^{-10}	5.7282×10^{-12}	8.9289×10^{-15}		

Table 6: Constants for the equation for correction factor J_r

A	B	C	D	E	F	G	H
−0.2477376	0.0129611	-1.127×10^{-5}	6.5427×10^{-8}	1.26556042	−0.0129666	2.5357×10^{-6}	1.5481×10^{-8}
I	**J**	**K**	**L**	**M**	**N**	**O**	**P**
−0.0481867	2.2364×10^{-4}	1.6928×10^{-5}	-1.708×10^{-7}	0.03112154	-2.758×10^{-4}	-6.979×10^{-6}	8.3062×10^{-8}

Correction Factor by Uneven Baffle Spacing at the Inlet and/or Outlet, J_s

This factor has an effect when exist a different baffle distribution at the inlet and/or outlet and along the tube bundle and it is computed by Equation (11):

$$J_s = \frac{(N_b - 1) + (l^*_{s,I})^{1-n''} + (l^*_{s,o})^{1-n''}}{(N_b - 1) + l^*_{s,I} + l^*_{s,o}} \qquad (11)$$

where $n'' = 0.6$ for turbulent flow and ($Re_C > 100$).

$n'' = 1/3$ for laminar flow and ($Re_C < 100$).

Correction Factor by Friction of an Ideal Bank Tubes f_i

The correction factor by friction in a triangular and rotated square set is determined in function of Re_C and the constants presented in Table 7 by the Equation (12).

$$f_i = \exp\{A + B\ln(Re_c) + C\ln(Re_c)^2 + D\ln(Re_c)^3 + E\ln(Re_c)^4 + F\ln(Re_c)^5 + G\ln(Re_c)^6 + H\ln(Re_c)^7 + I\ln(Re_c)^8 + J\ln(Re_c)^9\} \qquad (12)$$

A maximum error of 4.9% is result of comparing the original graph to that one obtained by Equation (12).

Pressure Loss Correction Factor by Tube-Baffle Leakage R_l

The introduction of this correction factor is due to the leakage of the tube bundle and it is computed by Equation (13) which is in terms of $S_1 = S_{sb}/(S_{sb} + S_{tb})$ and $S_2 = (S_{sb} + S_{tb})/S_m$.

$$R_l = \left[A + B(S_1) + C(S_1)^2 + D(S_1)^3\right] + \left[E + F(S_1) + G(S_1)^2 + H(S_1)^3\right]S_2 + \left[I + J(S_1) + K(S_1)^2 + L(S_1)^3\right](S_2)^2 + \left[M + N(S_1) + O(S_1)^2 + P(S_1)^3\right](S_2)^3 \qquad (13)$$

The constants required by the equations are established in Table 8. Comparing the original graph and that one obtained by Equation (13) is observed a maximum error of 4.5%.

Table 7: Constants for the equation for the correction factor f_i

Type arrangement, outside diameter and pitch between tubes			**A**	**B**	**C**	**D**	**E**
Arrangement	**Diameter**	**Pitch**					
Triangular	19.05 mm	23.8125 mm	4.15076	−0.675323	−0.254615	0.0590471	−0.00431448
Triangular	25.4 mm	31.75 mm	4.15076	−0.675323	−0.254615	0.0590471	−0.00431448
Rhombic	19.05 mm	25.4 mm	3.69311	−1.18662	0.281179	−0.136488	0.0280708
Triangular	19.05 mm	25.4 mm	3.85004	−0.609235	−0.278939	0.0630913	−0.00452036
Rhombic	25.4 mm	31.75 mm	3.97656	−0.796179	−0.1565	0.0319655	−0.00143937
Square	19.05 mm	25.4 mm	3.76203	−0.9323	−0.0827537	0.0678788	−0.0281861
Square	25.4 mm	31.75 mm	3.99352	0.768721	−0.32634	0.176985	−0.0326325
Type arrangement, outside diameter and pitch between tubes			**F**	**G**	**H**	**I**	**J**
Arrangement	**Diameter**	**Pitch**					
Triangular	19.05 mm	23.8125 mm	0.000102933	0	0	0	0
Triangular	25.4 mm	31.75 mm	0.000102933	0	0	0	0
Rhombic	19.05 mm	25.4 mm	−0.00237654	0.0000713989	0	0	0
Triangular	19.05 mm	25.4 mm	0.000103903	0	0	0	0
Rhombic	25.4 mm	31.75 mm	0.00547024	−0.000459287	0.000013822	0	0

Square	19.05 mm	25.4 mm	−0.00409219	0.00245958	0.0003604	0.0000227548	-5.37988×10^{-7}

Table 8: Constants for the equation for the correction factor R_l

Values of S_l	A	B	C	D	E	F	G	H
0 to 0.3	0.9930511	0.04163241	−0.07862612	0.01413662	−4.092978	−6.546514	4.516072	−2.525205
0.3 to 0.8	0.7499537	−0.4381533	0.2981431	0.1471556	−0.709333	0.0847061	1.053554	−4.215857
Values of S_l	**I**	**J**	**K**	**L**	**M**	**N**	**O**	**P**
0 to 0.3	15.6087	30.3289	−23.86598	14.08358	−23.22663	−41.67332	24.92778	−15.50438
0.3 to 0.8	0.3060496	1.27137	−12.85256	19.67211	−0.212271	−3.186438	18.491314	−23.16708

Pressure Loss Correction Factor by Recirculation Effect R_b

This correction factor is provoked by circulation currents in the heat exchanger and it is computed by Equation (14) which is in terms of F_{sbp} and $N_2 = N_{ss}/N_c$.

$$\begin{aligned} R_b = & \left[A + B(N_2) + C(N_2)^2 + D(N_2)^3 + E(N_2)^4\right] + \left[F + G(N_2) + H(N_2)^2 + I(N_2)^3 + J(N_2)^4\right] F_{sbp} \\ & + \left[K + L(N_2) + M(N_2)^2 + N(N_2)^3 + \tilde{N}(N_2)^4\right] F_{sbp}^2 + \left[O + P(N_2) + Q(N_2)^2 + R(N_2)^3 + S(N_2)^4\right] F_{sbp}^3 \\ & + \left[T + U(N_2) + V(N_2)^2 + W(N_2)^3 + X(N_2)^4\right] F_{sbp}^4 \end{aligned} \quad (14)$$

The constants demanded by the equation are defined in Table 9 and they are in terms of Re_c number. Comparing the original graph and that one obtained by Equation (14), the maximal error found is 1.3%.

Pressure Loss Correction Factor by Uneven Baffle Spacing at the Inlet and Outlet

This correction factor is proposed by the uneven baffle spacing at inlet and outlet of the heat exchanger and this factor is calculated by Equation (15).

$$R_s = \frac{1}{2}\left[\left(l_{s,I}^*\right)^{-n'} + \left(l_{s,o}^*\right)^{-n'}\right] \quad (15)$$

where $n' = 1.6$ for turbulent flow ($Re_c > 100$).

$n' = 1$ for laminar flow ($Re_c < 100$).

CALCULUS PROGRAM

Having the correction factors in an analytical form, it was developed a calculus program using the FORTRAN 77 language which will design quickly the shell and tubes heat exchanger through the improvement DELAWARE Method. This computing program is described next:

Flow Chart Program

The flow chart used for the shell and tubes heat exchanger is presented in Figure 2.

As from this flow chart, it was made the heat exchanger design program using the executable BELL since it runs in MS DOS platform.

CONCLUSIONS

This work has presented the improvements to the Delaware method for the shell and tube heat exchanger design. These improvements were made up not only of obtaining the correction factor equations which were only available in graphic forma but also of developing the computing program in Fortran 77 language.

Thus, a new and easy tool for the shell and tubes heat exchanger design is available which allows accomplishing the numerical computing in a quick form minimizing the errors of the graphical lecture. This system gives the opportunity to explore different design alternatives in order to find the optimal solution to each proposed problem.

Recent years, heat exchanger is often used for the request of technology. But the relevant design is not provided by the actual standards. This work presents the improvements to the Delaware method for the shell and tube heat exchanger design.

Table 9: Constants for the equation for the correction factor R_b

For	A	B	C	D	E	F	G	H	I
$Re_c > 100$	0.9999956	0.02427714	−0.2498571	0.87688954	−0.9389413	−3.7854133	42.7281376	−285.42201	841.836875
$Re_c < 100$	0.9989929	0.10536115	−0.8164392	2.13589149	−1.8323906	−4.7393642	60.6001915	−402.43318	1138.26259
For	**J**	**K**	**L**	**M**	**N**	**Ñ**	**O**	**P**	**Q**
$Re_c > 100$	−823.25082	6.9400004	−114.27156	831.289197	−2493.3366	2464.67857	−7.1165478	129.6448	−995.82525
$Re_c < 100$	−1075.7794	11.8151337	−238.94173	1739.20139	−4979.2121	4724.16651	−17.614757	423.093405	−3227.2538
For	**R**	**S**	**T**	**U**	**V**	**W**	**X**		
$Re_c > 100$	3071.9304	−3083.9014	3.21390213	−60.119316	480.616728	−1521.5693	1550.23124		
$Re_c < 100$	9354.34611	−8902.6809	11.6766845	−302.46692	2376.53333	−6980.2925	6687.42629		

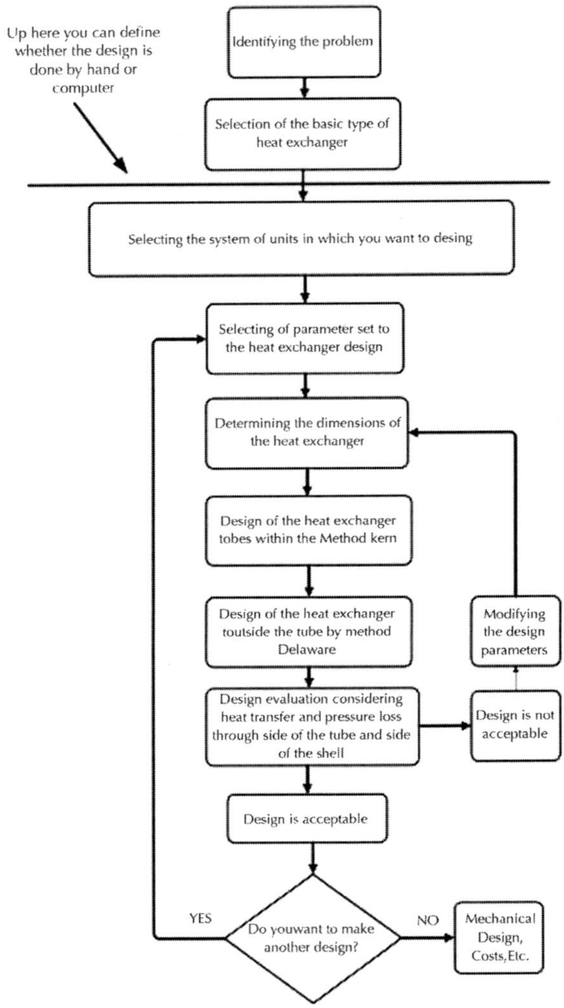

Figure 2: Flow chart program.

REFERENCES

1. Vengateson, U. (2010) Design of Multiple Shell and Tube Heat Exchangers in Series: E Shell and F Shell. Chemical Engineering Research and Design, 88, 725-736.http://dx.doi.org/10.1016/j.cherd.2009.10.005

2. Costa, A.L.H. and Queiroz, E.M. (2008) Design Optimization of Shell and tube Heat Exchangers. Applied Thermal Engineering, 28, 1798-1805.http://dx.doi.org/10.1016/j.applthermaleng.2007.11.009
3. Serna, M. and Jimenez, A. (2005) A Compact Formulation of Bell-Delaware Method for Heat Exchanger Desing and Optimation. Chemical Engineering Research and Desing, 83, 539-550. http://dx.doi.org/10.1205/cherd.03192
4. Shah, R.K., Dusan, P. and Sekulic, D.P. (2003) Fundamentals of Heat Exchanger Design. John Wiley & Sons, Hoboken. http://dx.doi.org/10.1002/9780470172605
5. Ayub, Z.H. (2005) A New Chart Method for Evaluating Single-Phase Shell Side Heat Transfer Coefficient in a Single Segmental Shell and Tube Heat Exchanger. Applied Thermal Engineering, 25, 2412-2420.http://dx.doi.org/10.1016/j.applthermaleng.2004.12.015
6. Leong, K.C., Toh, K.C. and Leong, Y.C. (1998) Shell and Tube Heat Exchanger Design Software for Educational Applications. International Journal of Engineering Education, 14, 217-224
7. León, A.R., Velazquez, M.T. and Diez, P.Q. (2011) The Design of Heat Exchangers. Engineering, 3, 911-920.
8. Castillo, R. (1999) Diseño Computacional de Intercambiadores de Calor de Coraza y Tubos por el Método Delaware. Tesis de Maestría, Instituto Politécnico Nacional, México.

Chapter 7

Heat Exchanger Specification: Coupling Design and Surface Performance Evaluation

Zhanying Zheng[a], Anthony M. Johnston[b], David F. Fletcher[a], and Brian S. Haynes[a]

[a]School of Chemical and Biomolecular Engineering, The University of Sydney, NSW 2006, Australia

[b]Optint Pty Ltd, Sydney, Australia

ABSTRACT

We develop an approach to obtaining the properties (flow area, path length and enclosed volume) of a heat exchanger surface that meets a design specification given simply in terms of the fluid properties, the flow rate, the allowable pressure drop, and the terminal fluid temperatures.

With the thermal performance of the surface to be designed expressed in terms of NTU (=| T|/ T_m for the side in question), we obtain a dimensional composite design specification which can be calculated simply from the inputs to the design. This group (on the left side of the equation below) is shown to depend only on the thermohydraulic properties of a candidate surface (right side below), thereby linking the flow Reynolds number and hydraulic mean diameter of the surface that satisfy the design:

$$\frac{\Delta p}{\mathrm{NTU}} \frac{2\rho}{\mu^2 \mathrm{Pr}^{2/3}} = \left(\frac{f}{j_H} Re^2 \right) d_h^{-2}$$

Since the flow area and path length can each also be expressed in terms of Re and d_h, the design is fixed once d_h is specified. The solutions are conveniently shown graphically such that the effects of using different passages sizes and different enhanced surfaces are revealed explicitly, without undertaking detailed design work. The method is ideal for early size and shape assessments, surface selection and optimisation of the design specification.

GRAPHICAL ABSTRACT

$$\frac{\Delta \rho}{NTU} \frac{2\rho}{\mu^2 Pr^{2/3}} = \left(\frac{f}{j_H} Re^2 \right) d_h^{-2}$$

Design Specification (duty,pressure drop, fiuid properties)	=	Surface performance characteristic $=f(\mathrm{Re}, d_h)$

INTRODUCTION

Over the last few decades, various heat transfer enhancement techniques have been proposed and are being used to improve the performance of heat exchangers, in terms of reduction of their size and cost, saving energy, etc. Webb and Kim (2005) compiled a comprehensive review of this topic, including over 9000 papers in their monograph. A recent review paper focussed on particular applications can be found inLeal et al. (2013).

With the continuing development of enhancement techniques, the question of how to evaluate these techniques has become crucial and has drawn considerable interest. Compared with plain surfaces, an enhanced surface often shows improved heat transfer but worse pressure-drop characteristics signified by the increases in both the Colburn factor, j_H, and the Fanning friction factor, f. It is therefore imperative that any evaluation method should enable the identification of the overall heat exchanger size reduction and shape change resulting from a particular enhancement technique. Crucial here is how the heat transfer and pressure drop characteristics are jointly evaluated and many approaches to this issue have been considered. Bergles et al. (1974) outlined various evaluation criteria for selecting and optimising heat exchanger passage geometries. Shah (1978) critically reviewed over thirty such evaluation methods in the 1970s and many more methods have been published since then.

A typical approach, still widely used, is that of London (1964) who proposed a term he called "surface flow area goodness factor" to compare the performance of heat exchange surfaces resulting from different enhancement techniques. This term can be expressed as

$$\frac{j_H}{f} = \frac{Nu\,Pr^{-1/3}}{f\,Re} = \frac{1}{A_X^2}\left[\frac{Pr^{2/3}}{2\rho}\frac{\dot{M}^2 NTU}{\Delta p}\right] \tag{1}$$

Here A_X denotes the cross-sectional flow area of the heat exchanger, while NTU and $_p$ are the number of heat transfer units required and pressure-drop allowed for the one side of the heat exchanger under consideration. Estimation of the value of the one-sided NTU formally requires the overall exchanger design to be finalised but there are circumstances in which the one-sided NTU may be estimated from the overall exchanger NTU, for example as $NTU_{one\ side} = 2NTU_{exchanger}$ for negligible wall resistance and equal heat transfer resistances on either side of the exchanger.

If the bracketed quantities of Eq. (1) are kept constant, then j_H/f varies inversely as A_x^2. Therefore, this evaluation method ranks enhancement techniques according to their ability to minimise the cross-sectional flow area required to satisfy a given design problem. It is not always apparent that this simple criterion is understood when applied in practice but recognition of the limitations in the flow area goodness factor has led to consideration of alternative ranking approaches.

London and Ferguson (1949) proposed a "volume goodness factor" evaluation method that incorporates the heat exchanger core volume. This approach can be demonstrated on a plot of heat transfer coefficient, h, as a function of the fluid pumping power per unit of heat transfer area, E:

$$h = \frac{k}{d_h} Nu = \frac{\hat{c}_p \mu}{Pr^{2/3}} \frac{1}{d_h} j_H Re \tag{2}$$

$$E = \frac{\dot{M} \Delta p}{\rho A_w} = \frac{\mu^3}{2\rho^2} \frac{1}{d_h^3} f\, Re^3 \tag{3}$$

For a given E, a greater value of h indicates a better performance because this corresponds to a smaller heat transfer area and therefore to a smaller heat exchanger volume. The dimensional nature of the parameters restricts the generality of this approach. However, Shah (1978) introduced modifications to this method by adding surface compactness to the equations to account for the effect of different ratios of heat transfer area to core volume, which improves the utility of the comparison.

Webb and co-workers have studied the ranking problem since the 1970s (Webb and Kim, 2005, Webb, 1981 and Webb and Eckert, 1972). They developed comparative assessments against reference surfaces as per the following two algebraic equations

$$\frac{(hA_w)_i}{(hA_w)_{ref}} = \frac{j_{H,i}}{j_{H,ref}} \frac{A_{w,i}}{A_{w,ref}} \frac{G_i}{G_{ref}} \tag{4}$$

$$\frac{P_i}{P_{ref}} = \frac{f_i}{f_{ref}} \frac{A_{w,i}}{A_{w,ref}} \left(\frac{G_i}{G_{ref}} \right)^3 \tag{5}$$

where the subscripts i and ref denote the values for enhanced and reference surfaces, respectively; G denotes the mass velocity; and P denotes the pumping power. Depending on different design requirements, Eqs. (4) and (5) can be simplified by fixing several parameters, thus enabling a certain parameter for the enhanced surface to be compared with that for the reference surface, e.g. for fixed hand A_w, P_i/P_{ref} can be derived by combining the above two equations, as below

$$\frac{P_i}{P_{ref}} = \frac{f_i/f_{ref}}{(j_{H,i}/j_{H,ref})^3} \tag{6}$$

In this way, the evaluation can be arranged to depend only on the values of j_H and f. In total, twelve similar cases of interest with various design objectives and fixed parameters were proposed to form a comprehensive Performance Evaluation Criteria group (Webb and Kim, 2005).

Soland et al. (1978) proposed an evaluation diagram of NTU per unit volume versus pumping power per unit volume. The fluid pumping power per unit volume is given by

$$\frac{P}{V} = \frac{\dot{M}\Delta p}{\rho V} = \frac{2\mu^3}{\rho^2} \frac{fRe^3}{d_h^4} \propto \frac{fRe^3}{d_h^4} \tag{7}$$

where V is the exchanger core volume and dh is the hydraulic mean diameter of the enhanced surface. The per unit volume NTU is given by

$$\frac{NTU}{V} = \frac{A_w h}{V\dot{M}\hat{c}_p} = \frac{4\mu}{\dot{M}Pr^{2/3}} \frac{j_H Re}{d_h^2} \propto \frac{j_H Re}{d_h^2} \tag{8}$$

For a particular value of fRe^3/d_h (P/V), a greater value of $j_H Re/d_h$ (NTU/V) corresponds to an enhanced surface with better performance. Sahiti et al. (2006) recently proposed a modification to this method by replacing NTU/V with $\dot{Q}/V$, where $\dot{Q}$ is the heat transfer rate. The modification resulted in a physically more direct comparison although with an apparent disadvantage because $\dot{Q}$ mostly needs to be identified through experimental measurements.

Fan et al. (2009) recently proposed a performance evaluation plot of Nui/Nu_{ref} versus fi/f_{ref} and provided multiple interpretations corresponding to different assumptions (identical pumping power, identical pressure-drop or identical flow rate), enabling the diagrammatical comparison of $\dot{Q}$ between the enhanced and reference surfaces.

There have also been numerous studies relying on the concept of entropy generation which was first proposed by Bejan, 1977 and Bejan, 1978 and then developed extensively by other researchers (Chen and Huang, 1988, Grazzini and Gori, 1988 and Hesselgreaves, 2000). These studies pointed out that entropy in a heat exchanger is generated in two ways, one associated with heat transfer and the other with frictional pressure-drop. Therefore, the surface with the least total entropy production gives the best thermodynamic performance, because it wastes the minimum amount of usable energy.

While many approaches have been considered, there are still gaps between the current evaluation methods and the practical heat exchanger design process. The initial task for the heat exchange designer is to meet the thermal specification, requiring the achievement of certain fluid temperature changes subject to certain pressure drop constraints, in a cost-effective configuration. The designer may also need to confirm that the design is compatible with any space and shape constraints which exist. Preferably, the heat exchanger specifications are not considered to be immutable, but fall within the overall scope of system optimisation, so that the specifications may be modified if it is to the benefit of the overall system costs. For example, the heat exchanger size might reduce substantially if higher pressure drops are allowed, or system energy consumption might be reduced if higher contact effectiveness is achieved.

It is the aim of this paper to characterise heat exchanger surface performance in such a way that the designer, and others unskilled in detailed design, can gain insight into the size and shape of an exchanger, and the relative merits of employing various surface types, without undertaking detailed design work. The required inputs are simply the fluid properties and flow rates, the allowable pressure drops, and the terminal fluid temperatures. The characterizations in this paper may therefore be of assistance in early size and shape assessments, surface selection and optimisation of the design specification.

This paper builds on previous work (Johnston and Haynes, 2002) in which the heat exchange specification for each side in the exchanger was characterised in terms of p/NTU. The heat transfer coefficient, h, was charted against p/NTU and p/L for particular fluids and heat exchanger surfaces. Thus, for a given thermal specification and fluid properties, h could be read from the charts and, from h and p, the flow path lengths could be determined. Since h characterises the success of the surface in reducing thermal resistance, it is used also in the calculation and ranking of heat transfer area and side volume.

In this paper we characterise the performance of heat exchange surfaces through graphical presentations that demonstrate the relationship to p/NTU of:

- Flow volume, V (rather than h);

- Flow length, L; and
- Flow area, A_X. (Whilst A_X is obtainable simply directly from V and L, its separate charting discloses some valuable insights.)

The above charts are "universalized" with regard to fluids. This is achieved by multiplying the axes (p/NTU, V and so on) by appropriate property groupings, as detailed in the following sections, such that charts developed for a heat exchange surface with a particular fluid may be used with any fluid. Whilst still dimensional (requiring specification of the hydraulic mean diameter for evaluation), the charts are fluid-independent, and can be considered as analogous to fluid-independent f and j_H charts, but more directly conveying a wealth of information regarding the size and shape of a heat exchange core, and supporting meaningful comparisons between various surfaces.

METHOD DEVELOPMENT AND EVALUATION PLOTS

Assumptions

The evaluation method here is developed in terms of a plate-type heat exchange surface in which all exchange area is assumed to be prime area. However, it is emphasised that the method is not limited to this specific type of surface. In finned heat exchangers, much of the area is in fact not prime area but, as discussed by Kays and London (1984), fin efficiencies can be calculated and used to derive an equivalent area that can be treated as 'prime' area – some iteration may be required to refine the fin efficiencies as the geometry emerges but this is not limiting to the method, especially for its application in preliminary design.

For total flow cross-sectional area, normal to the direction of flow, remaining constant along the flow path, the flow path length is given by

$$L = \frac{V}{A_X} \tag{9}$$

Any tortuosity or folding of flow paths will see L exceed the external length of the exchanger. There is no assumed relationship between the two sides in relation to the flow geometries; specifically, the flow lengths may differ. We do not consider this further here other than to point out that there are diverse techniques for bringing together fluids with different flow lengths, most obviously through simple crossflow contacting. However, while unmixed cross-flow heat exchange can approach 100% effectiveness, this is often accompanied by a substantial size penalty relative to pure counter-flow. Multipass crossflow performance can also approach that of counterflow under many circumstances.

The external dimensions of the exchanger core depend not only on the flow-path dimensions but also on method of construction, channel path and tortuosity, contacting scheme, etc. We do not consider these relationships further here.

In the basic design process, we seek to obtain the values of L, V and A_X from input parameters and characteristics of the surface. The exchange area Aw is obtained from A_X, L and the hydraulic mean diameter, d_h chosen for the surface – the process in itself is not difficult but there are many interacting variables and it is often not obvious what effects certain choices (for example of surface structure or of hydraulic diameter) have on the overall design under the constraint always of meeting the design specification.

Design Specifications

The input parameters are

- Fluid properties (ρ, μ, $\hat{c}_p$ and Pr) for the side in question.
- Mass flow rate on the side in question $\dot{M}$ ($=\rho u_m A_X$).
- Pressure-drop allowance for the side, Δp for incompressible

fluids, or Δp/p for compressible fluids (Kays and London, 1984).

- Thermal duty and/or combination of inlet and outlet fluid temperatures for both sides sufficient to enable the required thermal performance to be calculated in terms of NTU:

$$NTU_{exchanger} = \frac{UA_w}{\min\{\dot{M}\hat{c}_p\big|_h, \dot{M}\hat{c}_p\big|_c\}} = \frac{\max\{\Delta T_h, \Delta T_c\}}{\Delta T_m} \tag{10}$$

where ΔT_h and ΔT_c are the temperature ranges of the hot and cold fluid streams and ΔT_m is the mean temperature driving force for the chosen contacting scheme (such as counterflow or crossflow, for example).

For the purposes of decoupling the design of the two sides, we define the one-sided NTU as

$$NTU_{side} = \left.\frac{hA_w}{\dot{M}\hat{c}_p}\right|_{side} = \left.\frac{|\Delta T|}{\Delta T_m}\right|_{side} \tag{11}$$

where $\Delta T_{m,side}$ is that portion of the overall temperature driving force for the exchange that is attributable to the thermal resistance of the side under consideration. A reasonable first guess is generally not difficult – for example, in a gas–gas heat exchanger with similar fluids and surface configurations, $\Delta T_{m,side}$ would be reasonably similar on the two sides, whereas in a water-cooled gas cooler $\Delta T_{m,side}$ on the gas side might be 10 times that on the water side. In the latter case, considerations on the gas side primarily drive the exchanger design. For more accurate determinations, some iteration may be required after examination of the second side of the exchange – see also Kays and London (1984).

Basic Equations and Derivations

In the following, we consider explicitly how the thermohydraulic performance characteristics of a heat transfer surface (expressed here in terms of j_H and f) impact on the geometrical properties of the surface that meets a particular specification. The heat transfer performance in terms of the one-sided NTU is

$$NTU = \frac{hA_w}{\dot{M}\hat{c}_p} = \frac{h}{G\hat{c}_p}\frac{A_w}{A_X} = St\frac{A_w}{A_X} = \frac{j_H}{Pr^{2/3}}\frac{A_w}{A_X} \tag{12}$$

The factors f and j_H are often described as functions of Reynolds number using diagrams or correlations in the literature – typical examples can be found in Kays and London (1984). However, in deeply laminar, fully developed flows, for which the Nusselt number becomes independent of Prandtl number, j_H does show some dependence on Pr. On the other hand, since many enhanced surfaces rely on the repeated disruption of the wall boundary layer for their effectiveness, the Prandtl number dependence of j_H is frequently weak or negligible, even at low Reynolds number, as shown recently for repeating zigzag channels (Zheng et al., 2013).

There may also be a weak dependence on Pr when $Pr \lesssim 1$ but this is generally not well characterised except for straight pipe flow.

We note here that laminar flows in particular may exhibit significant variations in heat transfer coefficient along a flow path as hydraulic and thermal boundary layers are developed. We ignore these effects here and assume that the overall heat transfer performance is fully developed. With enhanced surfaces, the boundary layer is repeatedly disrupted and the flow is never fully developed locally but the total flow path generally incorporates many repeated units (zig-zags in printed circuit heat exchangers or fin structures in plate-fin exchangers). However, apart from the first few such units, heat transfer performance does not change from one unit to the next (Zheng et al., 2013). In practice, therefore, flow development effects can often be safely ignored unless there are very few repeating structures in the flow path. We therefore ignore entrance phenomena in the exposition of the method.

Eliminating the ratio of wall and cross-sectional areas from these equations yields

$$\frac{\Delta p}{\mathrm{NTU}} = \frac{f}{j_H}\frac{G^2}{2\rho}Pr^{2/3} \tag{13}$$

The quantity Δp/NTU represents the essential thermohydraulic requirement the surface must meet. The task then is to find the appropriate surface characteristics (j_H/f) and mass flux (G) that meet this requirement for the specified fluid (ρ, Pr). Here we note that j_H and f are generally expressed as functions only of the flow Reynolds number, and that G arises in the definition of the Reynolds number. We therefore write Eq. (13) as

$$\frac{\Delta p}{\mathrm{NTU}} = \frac{f}{j_H}\left(\frac{Re\cdot\mu}{d_h}\right)^2\frac{Pr^{2/3}}{2\rho} = \frac{f}{j_H}\left(\frac{Re}{d_h}\right)^2\left[\frac{\mu^2 Pr^{2/3}}{2\rho}\right] \tag{14}$$

The term in the square brackets on the r.h.s. entails only fluid properties which can be grouped with p/NTU to yield a "composite design specification" (l.h.s. in Eq. (15)) which must be matched by the "surface performance characteristic" (r.h.s.):

$$\frac{\Delta p}{\mathrm{NTU}}\frac{2\rho}{\mu^2 Pr^{2/3}} = \left(\frac{f}{j_H}Re^2\right)d_h^{-2} \tag{15}$$

The term in brackets in the surface performance characteristic is grouped because it is nominally a function only of the Reynolds number for a particular surface and will be independent of scale (i.e. of d_h) under conditions of geometrical similitude. This function is known in advance for a particular surface. The design task is therefore reduced to choosing appropriate values of Re and d_h for that particular surface. More specifically, if d_h is specified, then the value of Re needed to satisfy the composite design specification is determined.

This is shown explicitly in Fig. 1, in which Reynolds number is plotted against the surface performance characteristic $f Re^2 / j_H d_h^2$ for a straight pipe with d_h = 0.001 m at different Prandtl numbers. Here the values of j_H and f at Re ≤ 2300 are for fully-developed laminar flow (Shah and London, 1978) and those atRe > 2300 are from

Churchill (1977) and Gnielinski (2013). While the straight pipe is especially well studied,Fig. 1 is generally characteristic of all heat transfer surfaces: a noticeable dependence on Prandtl number in (deep) laminar flow but little or no such dependence otherwise. We also note that the appearance of multiple Re values satisfying the performance function at high Pr in the transitional flow regime is a function of the pronounced laminar–turbulent discontinuity in straight pipes and is unlikely to be evident in enhanced heat transfer in which turbulent-like behaviour is established even under laminar conditions (Zheng et al., 2013). Finally, we remind the reader that we do not account for entrance length effects here, which may lead to some error in the precise form of the graphs in this case (which, in terms of the discussion of entrance effects above, comprises just one repeating unit).

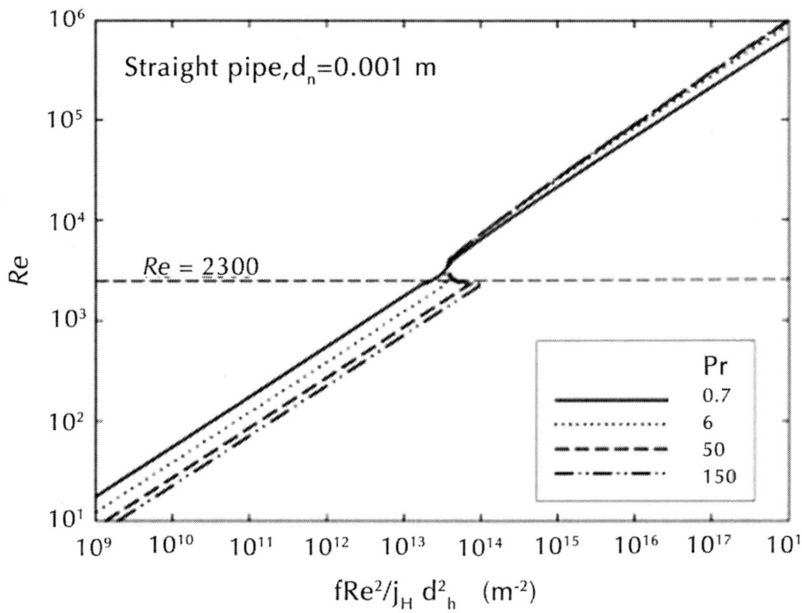

Figure 1: Reversus $(fRe^2/j_H d_h^2)$ for a straight pipe of $d_h = 0.001$ m at various Prandtl numbers.

Flow Area

The exchanger side flow area, A_X can be linked to the Reynolds number using the following dimensional expression

$$A_X \left[\frac{\mu}{\dot{M}}\right] = \frac{d_h}{Re} \tag{16}$$

The bracketed quantities of Eq. (16) belong to design inputs, so A_X can be determined for a given Reynolds number and channel dimension.

Eqs. (15) and (16) together provide the fundamentals for constructing a plot of $A_x\left[\mu/\dot{M}\right]$ as a function of the composite design specification, for use in evaluating the cross-sectional flow area of a core and, as shown in later examples, for comparing various surfaces on this basis. A specific surface is represented by a single curve on the plot (depending on the influence of the Prandtl number as pointed out above), as shown in Fig. 2 for a straight pipe of d_h = 0.001 m. While the data in this plot are the same as in Fig. 1, presentation in this way emphasises the use of the plot: design inputs determine the x-coordinate, from which the corresponding core cross-sectional flow area can be obtained from the appropriate curve. As thex-coordinate is a function only of the design specification and is the same regardless of which surface is chosen, curves for various enhanced surfaces can be presented on a single plot – those with higher locations correspond to exchangers with larger A_X. The exact values of A_X can be calculated from Eq. (16).

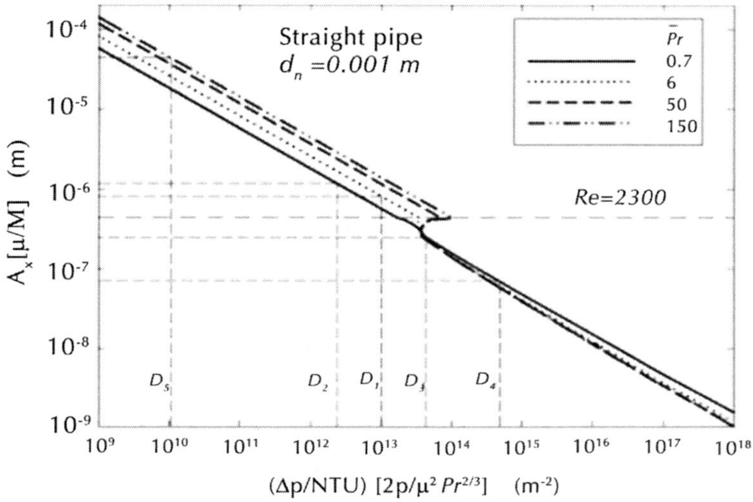

Figure 2: Flow area function versus composite design specification for heat exchanger using straight circular-section pipes with d_h = 0.001 m, showing weak dependence on Prandtl numbers in the range 0.7 < Pr < 150. The example design specifications D_1–D_5 are defined in Table 1 and results obtained from the chart are listed in Table 2.

To illustrate the range of x-coordinate for typical design problems, we consider five design specifications for water, air (at 1, 10 and 100 bar) and ethylene glycol, as given in detail in Table 1. These specifications cover a very wide range of the composite specification used as the x-coordinate in the evaluation chart, with the viscous liquid (ethylene glycol, D_5) having the lowest value and the high pressure gas (D_4) having the highest. Table 2 shows the values of AX for different design inputs obtained using the plot. It is a trivial matter then to extract other parameters of interest, such as velocities and Reynolds numbers, as also given in Table 2. For example, the number of tubular channels needed to meet each of the example design specifications ranges from about 1000 to nearly 10,000.

Table 1: Design input parameters for five example design problems

Design inputs	**D_1 (water)**	**D_2 (air, 1 bar)**	**D_3 (air, 10 bar)**	**D_4 (air, 100 bar)**	**D_5 (ethylene glycol)**
Δp (Pa)	50,000	500	1000	2000	100,000
NTU	3	2	2	3	3
ρ (kg m^{-3})	1000	1.2	11.3	113	1130
μ (Pa s)	0.001	0.000018	0.000018	0.00002	0.016
Pr	6	0.7	0.7	0.7	150
$\dot{M}$ (kg s^{-1})	1	0.1	0.1	1	1

Table 2: Exchanger geometry and flow parameters satisfying the individual design specifications in Table 1 using straight pipes of d_h = 0.001 m

	Design specifications				
	D1	D2	D3	D4	D5
$\frac{\Delta p}{NTU}\frac{2\rho}{\mu^2 P^{2/3}}$ (m−2)	1.01 × 1013	2.35 × 1012	4.42 × 1013	4.78 × 1014	1.04 × 1010
$A_X\left[\frac{\mu}{\dot{M}}\right]$ (m)	8.13 × 10−7	1.18 × 10−6	2.49 × 10−7	6.93 × 10−8	4.33 × 10−5

AX (m2)	8.13 × 10−4	6.56 × 10−3	1.38 × 10−3	3.47 × 10−3	2.71 × 10−3
N	1035	8347	1761	4412	3446
G (kg m−2 s−1)	1231	15	72	289	370
um (m s−1)	1.23	12.71	6.40	2.55	0.33
Re	1231	847	4016	14,430	23
$V\left[\frac{4\mu}{\dot{M}NTUPr^{2/3}}\right]$ (m2)	4.17 × 10−7	2.04 × 10−7	6.70 × 10−8	2.26 × 10−8	1.22 × 10−6
V (m3)	1.03 × 10−3	4.47 × 10−4	1.47 × 10−4	6.68 × 10−4	1.61 × 10−3
$L\left[\frac{4}{Pr^{2/3}NTU}\right]$ (m)	5.13 × 10−1	1.73 × 10−1	2.70 × 10−1	2.56 × 10−1	1.65 × 10−2
L (m)	1.270	0.068	0.106	0.151	0.349

Flow Volume

The flow volume, V, can be linked very simply to the definition of the hydraulic mean diameter of the surface:

$$V = \frac{1}{4} A_w d_h \tag{17}$$

Combining Eq. (17) with Eqs. (12) and (16), we obtain

$$V \left[\frac{4\mu}{\dot{M}\, NTU\, Pr^{2/3}} \right] = \frac{d_h^2}{Re j_H} \tag{18}$$

As previously, the bracketed quantities are design inputs, and the right-hand side is a function only of Reand dh (and to some extent also of Pr , as previously discussed). Therefore we can relate $V\left[4\mu / \dot{M}NTUPr^{2/3}\right]$ to $\Delta p / NTU(2\rho / \mu^2 Pr^{2/3})$ since the composite design specification can itself only be met for a given surface with specific values of Re and d_h. Fig. 3 shows the resulting plots (constructed using the same f and j_H values as in Fig. 1) for straight pipes of $d_h = 0.001$ m and for fluids with different Prandtl numbers. Once the curves for different surfaces are presented, their relative compactness can be evaluated simply by comparing their locations on the plot. As with Fig. 2, comparisons can be made only for the specific design inputs, so those with different Prandtl numbers as in Fig. 3 cannot be compared with each other.

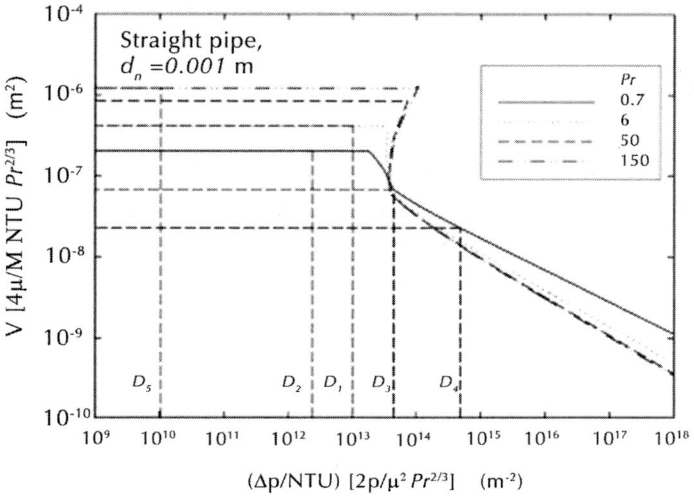

Figure 3: Enclosed volume function versus composite design specification for heat exchanger using straight circular-section pipes withd_h = 0.001 m, showing dependence on Prandtl numbers in the range 0.7 < Pr < 150. The example design specifications D_1–D_5 are defined in Table 1 and results obtained from the chart are listed in Table 2.

Extraction of the scaled enclosed volumes (Eq. (18)) for each of the design specifications D_1–D_5 in Table 1is also shown in Fig. 3, with results summarised in Table 2.

Flow Length, L

While the flow length, L can be evaluated from the separate determinations of volume, V and frontal area,A_X, it is instructive to present this graphically also. We divide Eq. (18) by Eq. (16) to obtain

$$L\left[\frac{4}{\mathrm{NTU}\,Pr^{2/3}}\right]=\frac{d_h}{j_H} \qquad (19)$$

where the bracketed quantities include all the design inputs and the right-hand side is a function only ofRe and dh (and to some extent also of Pr). Therefore we can relate L [4/NTU $Pr^{2/3}$] to the composite design specification which can only be met for a given surface with certain combinations of Re and d_h. As previously, we construct an evaluation plot to present scaled values of the design output, L in this case, for different surfaces. Fig. 4 shows the curves for a straight pipe of d_h = 0.001 m at different Prandtl numbers. Evaluations of the length functions for each of the design specifications D_1–D_5 in Table 1 are shown in the figure, with results summarised in Table 2.

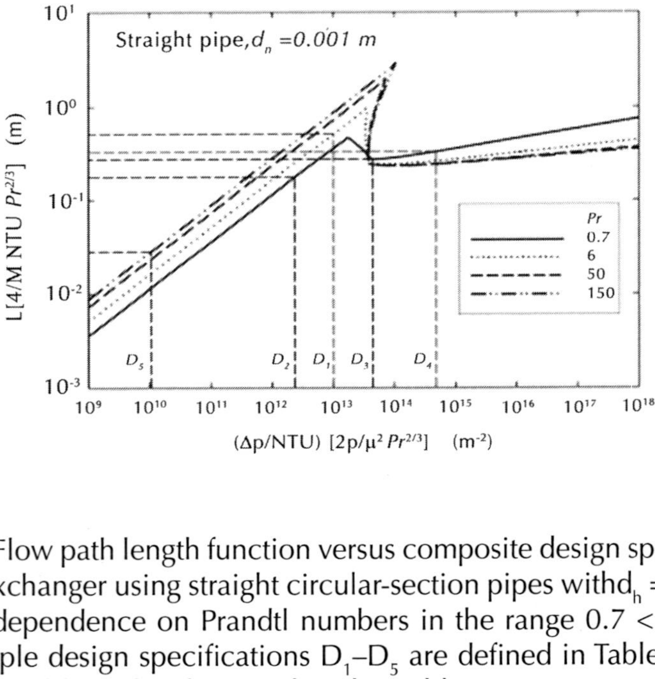

Figure 4: Flow path length function versus composite design specification for heat exchanger using straight circular-section pipes withd_h = 0.001 m, showing dependence on Prandtl numbers in the range 0.7 < Pr < 150. The example design specifications D_1–D_5 are defined in Table 1 and results obtained from the chart are listed in Table 2.

APPLICATIONS OF THE EVALUATION PLOTS

To this point, we have demonstrated the construction of the charts and shown their characteristic forms and interpretation. We now describe some particular applications.

Straight Pipe with Different Diameters

It is well-known that a small hydraulic diameter has the potential to reduce exchanger core size significantly. The new evaluation plots can be used to illustrate how core dimensional parameters vary as compactness is achieved.

Fig. 5 shows the evaluations of A_X, V and L for straight pipes of different hydraulic diameters carrying fluid with Pr = 6 (corresponding nominally to water). The charts are constructed using data for Fanning friction factor and Colburn factor described previously. The effect of channel size on relative core dimensions is given immediately by the respective locations of the curves – smaller channels clearly allow more compact exchangers (smaller V) to be used for the same duty, this being achieved primarily through a reduction in length rather than through any changes in flow area.

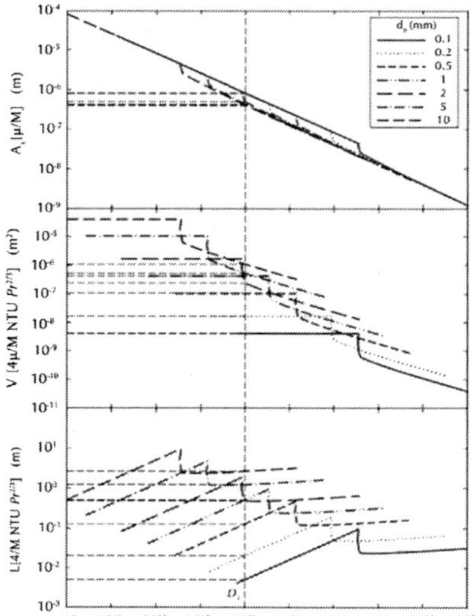

Figure 5: Evaluations of A_X, V and L for water flow (Pr = 6) in straight pipes with different hydraulic diameters. The exchanger geometry functions are

evaluated for each tube diameter for composite design specification corresponding to D_1 in Table 1.

We consider these effects in more detail for the D_1 design specification given in Table 1, for which the composite performance specification function value is 1.0×10^{13} m^{-2}, as shown in Fig. 5. The design points for the different choices of channel dimension, from 0.1 to 10 mm, are obtained directly from the graph and converted to core dimensions as summarised in Fig. 6.

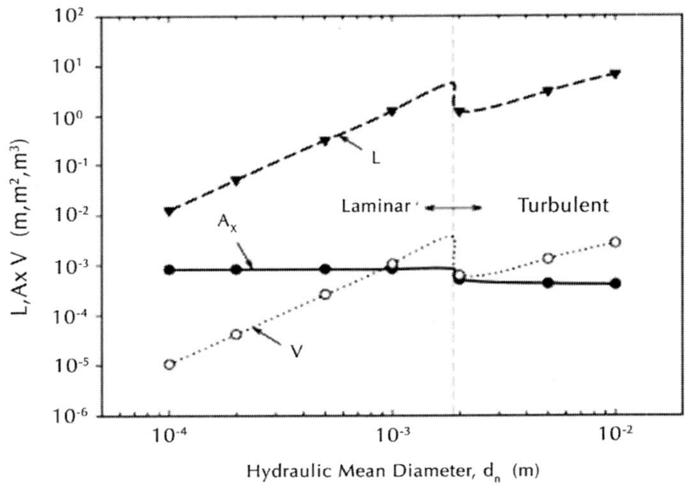

Figure 6: Evaluation of exchanger flow path dimensions for straight pipes satisfying the D_1 design condition given in Table 1.

An obvious implication of reducing channel dimension is that the Reynolds number at the design point is also reduced. The critical value of Re = 2300 corresponds to a hydraulic diameter $d_h \sim 2$ mm. Apart from a small increase in V in going through the turbulent-to-laminar transition, the overwhelming effect of reducing channel size is to reduce the enclosed fluid volume. In larger pipes, the design is for turbulent flow – in this region, $V \propto d_h^{1.1}$ so a 10-fold reduction in pipe diameter reduces the enclosed fluid volume by a factor ~12; in the laminar region, $V \propto d_h^2$ and the corresponding volume reduction would be 100-fold. Reductions in enclosed volume are achieved essentially only by the shortening of the flow path at the design point, the flow area remaining essentially constant whatever channel size is chosen. To satisfy the flow area requirement, the number of channels required increases as $\sim d_h^2$ across the whole range. In physical terms, the channel length goes from 6.8 m with 10 mm diameter tubes (5 in number) to 0.013 m with 0.1 mm diameter tubes (10^5 in number). Comparing $L/\sqrt{A_X}$ shows the approximate length/diameter ratio for the channels decreasing from 350 to 0.48 over this range of diameter reduction – at the upper end of the range, the geometry is typical of a tube bundle in a shell-and-tube exchanger, as expected. At the lower end, we would have a very short exchanger with a large face area – such an envelope may pose a number of difficulties, including stresses imposed by excessive axial temperature gradients, performance loss through longitudinal conduction, and a mismatch with the geometry requirements for the second fluid in the core, especially in a crossflow arrangement. Not surprisingly, therefore, theoretically achievable levels of heat transfer intensification in very fine passages may not always be practicable. A channel dimension around 1 mm is often seen.

The insensitivity of the flow area to changes in the hydraulic mean diameter is particularly striking. Kays and London (1984, p. 39) commented on the very limited range of velocities that could satisfy a given design problem when choosing among different compact surfaces but the effect is obviously worthy of closer attention. When constructing the flow area chart, we plot $y = d_h/Re$ versus $x = fRe^2/j_H d_h^2$. The form of the chart is therefore very simple:

$$y = \sqrt{\frac{f}{j_H}} x^{-1/2} \tag{20}$$

The flow area requirement for any design problem therefore depends only on the flow area goodness factor j_H/f, as pointed out by London (1964). Beyond that, however, we can say the there is no direct influence of channel diameter on the design flow area other than through the variation of goodness factor with Reynolds number. In turbulent flow, we have $j_H/f \sim 0.5$, according to the basic Chilton–Colburn analogy. In fully-developed laminar flow, we have Nu = constant, ~4, so $j_H = 4Re^{-1} \cdot Pr^{-1/3}$; also $f = 16Re^{-1}$ so $j_H/f \sim 0.25Pr^{-1/3}$. Therefore, in either the laminar or turbulent cases, the area goodness factor is constant and formally independent of the hydraulic diameter. The laminar and turbulent lines on the graph for flow area are parallel, with slope –1/2. The offset between the two is given by

$$\frac{Ax|_{laminar}}{Ax|_{turbulent}} = \frac{\sqrt{j_H/f|_{turbulent}}}{\sqrt{j_H/f|_{laminar}}} = \sqrt{2} \cdot Pr^{1/6} \quad (21)$$

The difference between laminar and turbulent flow areas that meet the design specification therefore depends only weakly on the value of Pr. Overall, the flow area that satisfies a given design problem is very tightly constrained, hardly varying, if at all, over many orders of magnitude in channel dimension.

Evaluation of Various Enhanced Surfaces

The intended use of the plots in this study is to provide evaluations for various enhanced surfaces. We provide here some illustrative examples.

Three types of surfaces designed for heat transfer enhancement in the literature are selected. A wavy-fin surface is presented in Fig. 7(a), which has been studied experimentally in Kays and London (1984); Fig. 7(b) presents an offset-strip fin surface for which Fanning friction and Colburn factors can be found in Kim et al. (2011); Fig. 7(c) shows a wire coil structure inserted in a smooth tube (Garcia et al., 2005). Surfaces with typical geometrical parameters for the above three designs are evaluated together with a straight pipe geometry, all with dh = 0.001 m. Results are shown across a wide range of composite performance specifications in Fig. 8 where the design specification D_1 given in Table 1 is identified specifically.

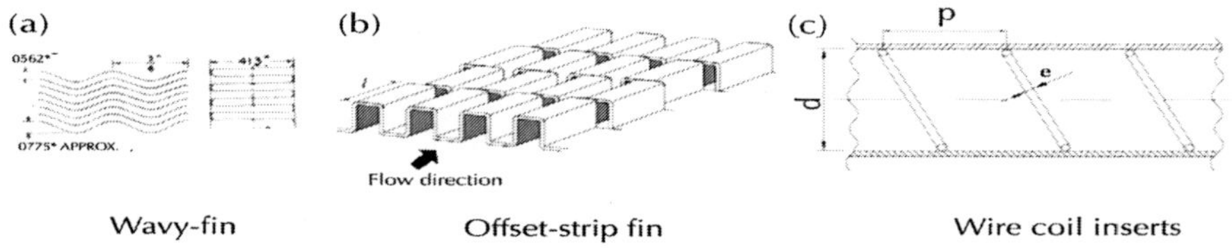

Figure 7: Schematics of the (a) wavy-fin surface (Kays and London, 1984), (b) offset-strip fin surface (Kim et al., 2011) and (c) wire coil inserts in smooth tube (Garcia et al., 2005).

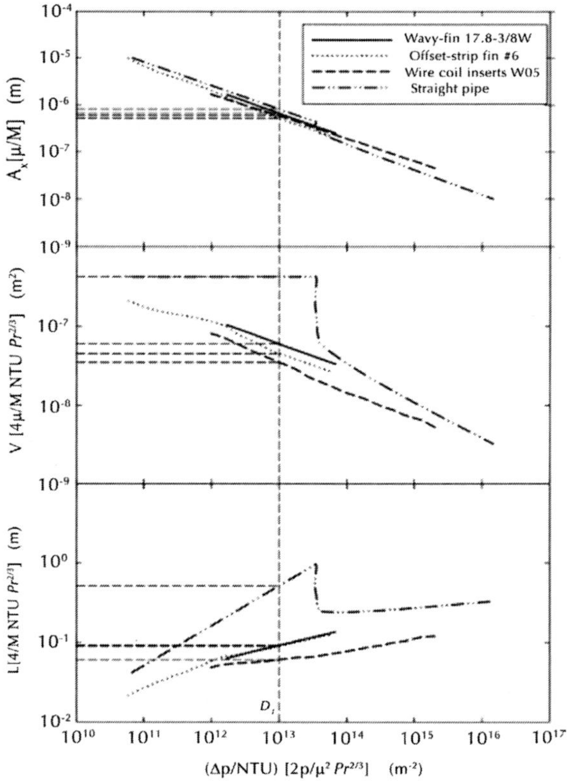

Figure 8: Evaluations of area, volume and length functions for different enhanced surfaces. Results for straight pipes are also shown, for compari-

son. All surfaces are assumed to have d_h = 0.001 m. The design point D_1 is defined in Table 1.

The design flow area is seen in Fig. 8(a) to be very nearly the same in all cases, as expected from the forgoing discussion. The line for the straight pipe shows the usual discontinuity at the laminar–turbulent transition but the other surfaces do not display such behaviour because secondary flows and unsteadiness are induced in flows over these surfaces, producing turbulent-like behaviour even at relatively low values ofRe. Here it should be recalled that, for constant hydraulic mean diameter, nearly constant flow area means that the different surfaces will have very similar Reynolds numbers at their respective design points.

In Fig. 8(b), the profound intensification achievable under laminar flow conditions with the enhanced surfaces is evident, particularly as the Reynolds number approaches the laminar–turbulent transition. Among the enhanced surfaces, there are still clear differences with wire coil inserts offering the smallest enclosed volume. The onset of turbulence in the straight pipe dramatically reduces the internal volume advantage offered by the enhanced surfaces, as expected. Also, at very low Reynolds numbers, the turbulisation mechanisms employed in the enhanced surfaces become ineffective and the internal volumes for the different surfaces all begin to converge again.

The channel length behaviour (Fig. 8(c)) is essentially the inverse of the enclosed volume trends and is not discussed further here.

The D_1 design inputs group is presented here to demonstrate the application of the charts to the task of surface selection and overall design. Based on the values extracted from the charts, the flow area for the three enhanced surfaces are all about 0.7 of that for a straight pipe. The internal volume V for wire coil inserts, offset-strip fin and wavy-fin surfaces are 0.08, 0.11 and 0.14 times that for a straight pipe, respectively. Relative path lengths are 0.18 for wavy-fin and offset-strip fin surfaces and 0.11 wire coil inserts. In order to make these comparisons on the basis of the heat exchanger core rather than the enclosed volumes, manufacturing considerations need to be taken into account. Beyond that, the matching of the design for the second fluid surface has to be undertaken – this entails judicious choices regarding surfaces, flow routing and channel diameters for both fluids.

Rating and Response to Design Variations

The charts developed are most directly suited to the basic design process of responding to a fixed specification. However, as discussed in Section 1, there are circumstances that require the designer to understand the sensitivity of the design to variations in the design specification. We consider here such effects in relation to the enhanced surfaces in Fig. 8.

An increase in the pressure drop allowance (at constant d_h) simply increases the value of the composite design specification. As a result, the flow area A_X decreases as $\sim\Delta p^{0.5}$ (Eq. (20)). On average for the three enhanced surfaces, the enclosed volume also decreases, $\sim\Delta p^{-0.37}$, while the flow path length increases marginally, $\sim\Delta p^{0.13}$. Clearly, only modest levels of intensification are achievable through greater pressure drop – the use of finer passages can give much more profound intensification (Fig. 6) without any pressure-drop penalty.

A higher effectiveness of heat exchange demands more NTU and a decrease in the value of the composite design specification. The effect on flow area is immediately apparent, $AX \propto NTU^{0.5}$. The effect on the enclosed volume and flow path length is not just the inverse of effect of pressure drop becauseNTU appears in the evaluation groups for these quantities – overall, $\Delta \log V \propto (0.37 + 1) \Delta \log NTU = 1.37 \Delta \log NTU$;; similarly $\Delta \log L \propto 0.87 \Delta \log NTU$. The higherNTU requirement entails a somewhat larger exchanger but this could be offset by choosing a higher pressure drop or a slightly reduced value of d_h.

In a rating problem, it may be desired to determine the thermal performance and pressure drop for a given flow rate, $\dot{M}$, when A_X, V and L are known. In such cases, we can form the area group ($A_x\mu/\dot{M}$) immediately and so fix the value of the composite design specification. It is a simple matter then to read off the length function ($4L/[NTU\ Pr^{2/3}]$) and hence evaluate NTU and finally Δp As a variant on this rating problem, we may seek to find the flow rate that can achieve a desired thermal performance – now the NTU is known so the length group can be evaluated to identify the values of the composite design specification and, from the area group, $\dot{M}$ and ultimately Δp.

CONCLUSIONS

The evaluation of heat exchanger surfaces is often made in isolation of actual design constraints. These constraints are readily expressed in term of the pressure drop allowances Δp and the thermal NTUrequirements for each side of the exchanger. The ratio Δp/NTU encapsulates the overall task for one side and is the underlying justification for the widely used comparison of heat transfer factor j_H and friction factorf as a basis for comparing the axial flow areas for different exchangers satisfying a given design. However, the ratio j_H/f says nothing about the overall size of the exchanger and is in itself essentially meaningless as a method of comparing the suitability of one surface versus another for a specific design.

We have developed a composite design specification that can be calculated from the design inputs and equated to a dimensional surface performance characteristic that is typically a function only of the flow Reynolds number, Re and hydraulic mean diameter, d_h of the surface. At the same time we derive flow area and length (or volume) functions that must be met to satisfy the design – these functions also depend only on Re and d_h. On this basis we construct plots that, for a chosen value of d_h, enable the flow passage geometry (L, A_x and V) to be determined. We are therefore able to compare directly different surface types with different d_h and thereby determine the performance of the surfaces directly in the context of the specific design task at hand. This formulation and the resulting charts are also very useful for direct solution of rating problems. The axial flow area found for different surfaces over a very wide range of passage sizes is very nearly invariant, in part because of the narrow range of j_H/f that arises. As a simple rule-of-thumb, it can be said that the flow area for a highly compact, enhanced-surface heat exchanger is very nearly the same as for the tube-side of a conventional shell-and-tube exchanger satisfying the same design specification.

A corollary of the invariance of the flow area is that intensification through the use of finer passages is manifest principally in reductions in the flow path length. Very fine passages may give rise to such short path lengths as to create fabrication difficulties, not to mention the potential loss of thermal performance through axial wall conduction. Enhanced surfaces also produce compactness mainly through reductions in the flow path length. The charts emphasise how enhanced surfaces achieve greatest intensification at Reynolds numbers in the vicinity of the laminar–turbulent transition for tube flow – the turbulisation and boundary-layer disruption promoted by enhanced surfaces is ineffective in deep laminar flows and redundant in highly turbulent flows.

A comparison of three enhanced surfaces reveals broad similarities and specific differences in their overall behaviour.

While the emphasis in this paper has been on graphical presentation, it is clear that the methodology described here is readily encoded for computational evaluation. For the basic design task, in which the composite design specification is established, the only free variable is the hydraulic mean diameter d_h, from which A_X and V (or L) can be determined explicitly for a given surface.

ACKNOWLEDGEMENTS

Zhanying Zheng acknowledges the financial support from Chinese Scholarship Council and the University of Sydney for his Ph.D. study. Discussions with Dr. Sharon Leung and Chris Jazrawi are also gratefully acknowledged.

REFERENCES

1. Bejan, A., 1977. The concept of irreversibility in heat exchangerdesign: counterflow heat exchangers for gas-to-gasapplications. J. Heat Transfer 99 (3), 374–380.
2. Bejan, A., 1978. General criterion for rating heat-exchangerperformance. Int. J. Heat Mass Transfer 21 (5), 655–658.
3. Bergles, A.E., Junkhan, G.H., Bunn, R.L., 1974. Performance criteriafor cooling systems on agricultural and industrial machines.Iowa Univ. Prog. Rep. 1.
4. Chen, B.H., Huang, W.H., 1988. Second-law analysis forheat-transfer enhancement on a rib-type turbulencepromoter. Energy 13 (2), 167–175.
5. Churchill, S.W., 1977. Friction-factor equation spans all fluid-flowregimes. Chem. Eng. 84 (24), 91–92.
6. Fan, J.F., Ding, W.K., Zhang, J.F., He, Y.L., Tao, W.Q., 2009. Aperformance evaluation plot of enhanced heat transfertechniques oriented for energy-saving. Int. J. Heat MassTransfer 52 (1–2), 33–44.
7. Garcia, A., Vicente, P.G., Viedma, A., 2005. Experimental study ofheat transfer enhancement with wire coil inserts inlaminar–transition–turbulent regimes at different Prandtlnumbers. Int. J. Heat Mass Transfer 48 (21–22), 4640–4651.
8. Gnielinski, V., 2013. On heat transfer in tubes. Int. J. Heat MassTransfer 63, 134–140.
9. Grazzini, G., Gori, F., 1988. Entropy parameters for heat exchangerdesign. Int. J. Heat Mass Transfer 31 (12), 2547–2554.
10. Hesselgreaves, J.E., 2000. Rationalisation of second law analysis ofheat exchangers. Int. J. Heat Mass Transfer 43 (22), 4189–4204.
11. Johnston, A.M., Haynes, B.S., 2002. Design considerations forcompact heat exchangers. In: Proceedings of

InternationalSymposium on Compact Heat Exchangers, Grenoble, France.

12. Kays, W.M., London, A.L., 1984. Compact Heat Exchangers,McGraw-Hill, New York.
13. Kim, M.S., Lee, J., Yook, S.J., Lee, K.S., 2011. Correlations andoptimization of a heat exchanger with offset-strip fins. Int. J.Heat Mass Transfer 54 (9–10), 2073–2079.
14. Leal, L., Miscevic, M., Lavieille, P., Amokrane, M., Pigache, F.,Topin, F., Nogarede, B., Tadrist, L., 2013. An overview of heattransfer enhancement methods and new perspectives: focuson active methods using electroactive materials. Int. J. HeatMass Transfer 61, 505–524.
15. London, A.L., 1964. Compact heat exchanger: Part 2. Surfacegeometry. Mech. Eng. 86, 31–34.
16. London, A.L., Ferguson, C.K., 1949. Test results ofhigh-performance heat exchanger surfaces used in aircraftintercoolers and their significance for gas-turbine regeneratordesign. Trans. ASME 71, 17–26.
17. Sahiti, N., Durst, F., Dewan, A., 2006. Strategy for selection ofelements for heat transfer enhancement. Int. J. Heat MassTransfer 49 (19–20), 3392–3400.
18. Shah, R.K., 1978. Compact heat exchanger surface selectionmethods. Heat Transfer 4, 193–199.
19. Shah, R.K., London, A.L., 1978. Laminar flow forced convection inducts: a source book for compact heat exchanger analyticaldata. Academic Press, New York.
20. Soland, J.G., Mack, W.M., Rohsenow, W.M., 1978. Performanceranking of plate-fin heat-exchanger surfaces. J. Heat Transfer:Trans. ASME 100 (3), 514–519.
21. Webb, R.L., 1981. Performance evaluation criteria for use ofenhanced heat-transfer surfaces in heat-exchanger design. Int. J. Heat Mass Transfer 24 (4), 715–726.
22. Webb, R.L., Eckert, E.R.G., 1972. Application of rough surfaces toheat-exchanger design. Int. J. Heat Mass Transfer 15 (9), 1647–1658.

23. Webb, R.L., Kim, N.H., 2005. Principles of Enhanced Heat Transfer,second ed. Taylor & Francis, New York.

24. Zheng, Z., Fletcher, D.F., Haynes, B.S., 2013. Laminar heat transfersimulations for periodic zigzag semicircular channels: chaoticadvection and geometric effects. Int. J. Heat Mass Transfer 62(1), 391–401.

Chapter 8

The Effect of Flow Arrangement on the Pressure Drop of Plate Heat Exchangers

Raquel Y. Miura[a], Flavio C.C. Galeazzo[b], Carmen C. Tadini[a], and Jorge A.W. Gut[a]

[a]Department of Chemical Engineering, Escola Politécnica, University of São Paulo, ,São Paulo, SP05424-970, Brazil

[b]Division of Combustion Technology, Engler-Bunte-Institute, University of Karlsruhe, Germany

ABSTRACT

For the optimal design of plate heat exchangers (PHEs), an accurate thermal–hydraulic model that takes into account the effect of the flow arrangement on the heat load and pressure drop is necessary. In the present study, the effect of the flow arrangement on the pressure drop of a PHE is investigated. Thirty two different arrangements were experimentally tested using a laboratory scale PHE with flat plates. The experimental data was used for (a) determination of an empirical correlation for the effect of the number of passes and number of flow channels per pass on the pressure drop; (b) validation of a friction factor model through parameter estimation; and (c) comparison with the simulation results obtained with a CFD (computational fluid dynamics) model of the PHE. All three approaches resulted in a good agreement between experimental and predicted values of pressure drop. Moreover, the CFD model is used for evaluating the flow maldistribution in a PHE with two channels per pass.

INTRODUCTION

Plate heat exchangers (PHEs) were originally designed for hygienic applications in the food and pharmaceutical industries. However, in the last decades, PHEs have been extensively used in the power and process industries as well. In addition to the ease of sanitation and maintenance, the main advantages of PHEs are the compactness, flexibility, and high thermal effectiveness. Basically, the PHE consists of a compressed pack of thin metal plates, where each plate has a sealing gasket around the perimeter and can have up to four corner orifices (ports). The gaskets are designed so that hot and cold fluids flow through alternate flow channels without mixing. Heat is exchanged between neighbor channels through the metal plates, which have a corrugated (wavy) surface for improving turbulence and mixing. The alignment of the plate orifices forms continuous ducts (manifolds) inside the plate pack that distribute the fluids into the narrow channels between plates (Shah and Focke, 1988; Kakaç and Liu, 2002).

A very large number of flow arrangements is possible, depending on the perforations of the plates and gasket design. The fluid entering the plate pack at its top or bottom is split into N parallel flow channels and is collected in the duct on the opposite side. This group of channels in which the flow is in the same direction is denominated "pass". Both hot and cold sides of the PHE consist of a series of passes, which usually have a constant number of channels per pass. The flow arrangement of a PHE is often presented as $P_{hot} \times N_{hot}/P_{cold} \times N_{cold}$, where *P* is the number of passes and *N* is the number of channels per pass for hot or cold sides. The number of channels in the hot or cold sides of the PHE is given by P·N, the total number of channels in the PHE is thus defined as $N_C = P_{hot} \cdot N_{hot} + P_{cold} \cdot N_{cold}$ and the number of thermal plates is given by N_C-1. Fig. 1 shows a few examples of possible configurations for a PHE with seven channels. The configuration consists of the flow arrangement and the location of the inlet and outlet connections of hot and cold fluids.

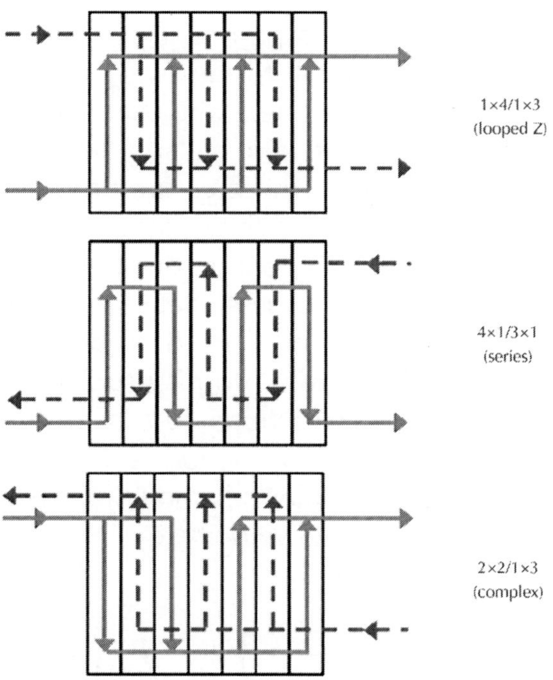

Figure 1: Some possible configurations for a PHE with seven flow channels (six thermal plates).

The thermal–hydraulic performance of a PHE is strongly influenced by the geometry of the plate corrugation and by the exchanger configuration. In order to improve the turbulence inside the channels the angle between the flow direction and the corrugation orientation can be increased, or the number of channels per pass can be reduced for increasing the velocity of the fluid inside the channels. However, the turbulence not only increases the convective heat transfer coefficient but also increases the pressure drop. Consequently, heat transfer and pressure drop are coupled and an accurate thermal–hydraulic model is required for the design of PHEs for meeting minimum heat load and maximum pressure drop constraints. Moreover, the model must properly take into account the effect of the PHE configuration on its performance for the optimal design of the flow arrangement (Gut and Pinto, 2004 and Gut et al., 2004).

For the design of PHEs, the pumping power is an important constraint, which is proportional to the pressure drop. Moreover, PHEs have pressure limitations due to the extensive use of gaskets; consequently, the maximum pressure is also a design constraint that depends on the pressure drop. The pressure drop associated with the plate pack consists mainly of three components: (1) pressure drop within the channels due to friction and flow contraction and expansion; (2) pressure drop associated with the distribution ducts inside the PHE; and (3) pressure drop due to an elevation change (static head). The pressure drop is usually evaluated using an analogy to the single-phase frictional loss inside tubes, as shown in Eq. (1), where *f* is the Fanning friction factor, *L* is the channel length, $D_e = 2 \cdot d/\Phi$ is the equivalent diameter for the PHE channel, $v_c = Q/(N \cdot A_c)$ is the mean channel velocity, $v_p = Q/A_p$ is the mean port velocity (Shah and Focke, 1988):

$$\Delta P_{PHE} = \frac{4 \cdot f \cdot L \cdot P}{D_e} \cdot \left(\frac{\rho \cdot v_c^2}{2} \right) + \lambda \cdot P \cdot \left(\frac{\rho \cdot v_p^2}{2} \right) \tag{1}$$

The first term in the right-hand side of Eq. (1) is associated with the frictional loss, whereas the last term evaluates the total losses within the distribution ducts (static head is not included in Eq. (1)). Empirically, the duct pressure drop is assumed proportional to the inlet velocity head through the plate port. Saunders (1988) recommends using the proportionality coefficient λ=1.3, Shah and Focke (1988) suggest λ=1.5, and Kakaç and Liu (2002) propose λ=1.4. According to Shah and Focke (1988), the pressure drop in the port ducts should be kept as low as possible by enlarging the pipe and plate port diameters because this pressure drop does not contribute with the heat transfer. In well-designed PHEs, this component should be below 10% of the exchanger pressure drop.

In the present study the effect of the flow arrangement on the pressure drop of a PHE is investigated. Pressure drop data was obtained from a laboratory scale PHE with varying flow rate using a large number of different flow arrangements. The experimental data is used for (a) determination of the influence of the number of passes and number of channels per pass on the pressure drop; (b) validation of the model in Eq.(1) through parameter estimation; and (c) comparison with the simulation results obtained from a CFD (computational fluid dynamics) model of the PHE.

EXPERIMENTAL PROCEDURE

The PHE used for this study was an FT-43 laboratory plate pasteurizer (Armfield, Hampshire, UK), which has 12 × 8cm stainless steel flat plates without corrugation with silicon gaskets (see Fig. 2); the plate main dimensions are presented in Table 1. Fig. 3 presents a schematic of the experimental setup: a centrifugal pump circulated distilled water at ambient temperature in a closed circuit that included both sides of the PHE. Flow rate (Q) was monitored by a rotameter and controlled with a needle valve (maximum flow rate was 1.45 L/min). Temperature (T) was measured with a thermocouple at the tank inlet and the pressure drop (ΔP_{exp}) at one side of the PHE sides was monitored with a differential pressure sensor (Foxboro, MA, USA), as indicated in Fig. 3.

Figure 2: PHE (left) and opened plate pack showing plates and gaskets of a 1×2/1×2 flow arrangement (right).

Table 1: Main plate dimensions of the FT-43 PHE

Dimension	**Value**
Channel length, between port centers (mm)	90
Channel width (mm)	60
Plate thickness (mm)	1.0
Mean channel gap (mm)	1.5
Equivalent diameter (mm)	3.0
Port diameter (mm)	6.5
Heat transfer area (m²/plate)	0.0050

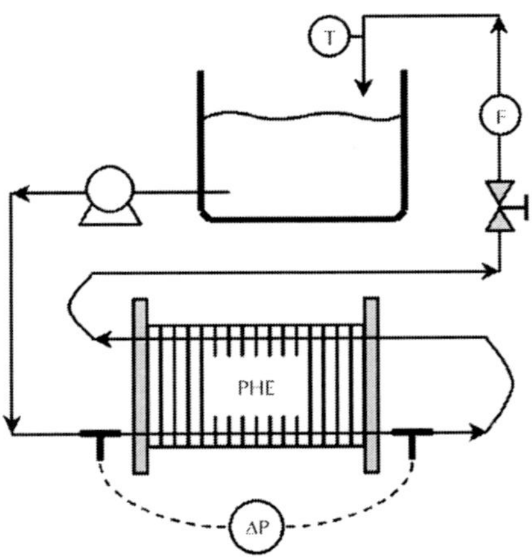

Figure 3: Schematic of the experimental setup for pressure drop measurement.

Thirty two different flow arrangements were tested for the monitored side of the PHE, which are (P×N): 0×0, 1×1, 1×2, 1×3, 1×4, 1×5, 1×6, 1×7, 2×1, 2×2, 2×3, 2×4, 3×1, 3×2, 3×3, 4×1, 4×2, 5×1, 5×2, 6×1, 7×1, 8×1, 9×1, 10×1, 11×1, 12×1, 13×1, 14×1, 15×1, 16×1, 17×1, and 18×1 (the number of arrangements was limited by the available types of plates and gaskets). The largest number of channels tested was 18. The flow arrangement 0×0 consisted of a single gasket, with the objective of obtaining the pressure drop of the pipeline between measuring points, denominated ΔP_{pipe}.

For each experimental run, first the plate pack was configured and aligned for the desired flow arrangement, the PHE was closed, the compression bolts tightened and water was circulated with varying flow rate for removing the air bubbles trapped inside the system. Experiments were then conducted with increasing and decreasing flow rates to reduce the hysteresis effects. The step change in flow rate was 0.05 L/min and ΔP and *T* were measured each time. The uncertainties for measurement of ΔP, *Q*, and *T* were assumed to be 5%, 2%, and 1%, respectively. Because of the small dimensions of this PHE, the pressure drop is very low (readings were between 1 and 7800 Pa) and it showed to be very sensitive to the presence of air bubbles, to the compression applied to the plate pack (Leuliet et al., 1990), and to the alignment of the gaskets and plates. Several runs were repeated for the same conditions to check the repeatability of the results, thus identifying the main sources of variation. For minimizing these effects on the final results, at least five independent runs were conducted for each individual flow arrangement for identification of its average behavior.

ANALYSIS OF RESULTS AND DISCUSSION

Correlation for the Effect of the Flow Rate

The first step of data treatment was to identify the effect of the flow rate on the pressure drop. A log–log plot of measured pressure drop (Pa) as a function of flow rate (L/min) showed a clear linear dependence, as was also verified by Leuliet et al. (1990) for a PHE with washboard plates. By comparison of the results obtained for any given flow arrangement, the deviating runs were discarded and three or four runs were selected to represent the average behavior of each one of the 32 flow arrangements. The number of valid runs was 103 with 1918 experimental points. The mean temperature was 23±2°C. The measured experimental pressure drop, ΔP_{exp}, consists of three components: the pressure drop of the PHE (ΔP_{PHE}), the pressure drop due to the pipeline between measuring points (ΔP_{pipe}) and the static pressure drop (ΔP_0); thus, the pressure drop of the PHE can be obtained from Eq. (2), if the components ΔP_{pipe} and ΔP_0 are known.

$$\Delta P_{PHE}=\Delta P_{exp}-\Delta P_{pipe}-\Delta P_0 \qquad (2)$$

The pressure drop data of each valid run were fitted using Eq. (3a) by adjusting parameters ΔP_0 (Pa), α(Pa (L/min)$^{-\beta}$) and β (dimensionless). The runs for flow arrangement 0×0 were treated similarly using Eq. (3b):

$$\Delta P_{exp}=\Delta P_0+\alpha\cdot Q^{\beta} \qquad (3a)$$

$$\Delta P_{pipe}=\Delta P_0+\alpha_{pipe}\cdot Q^{\beta} \qquad (3b)$$

The effect of *P* and *N* on parameter β is presented in Fig. 4. No apparent correlation can be observed between *P* and β. The effect of *N* on parameter β was also assumed insignificant, since more than half of the runs have N=1 or 2, and only a few (10%) have N ≥5. Therefore, the mean value $\beta=1.76\pm0.08$was assumed for this work. For each flow arrangement, all experimental pressure drop data (subtracting the respective static pressures, ΔP_0) were correlated with $Q^{1.76}$ for determining parameter α. An example is shown in Fig. 5: experimental data from three independent runs with flow arrangement 11×1were used for the determination of $\alpha=8452$. The mean absolute error on the determination of pressure drop using Eqs. (3a) and (3b) is 7%.

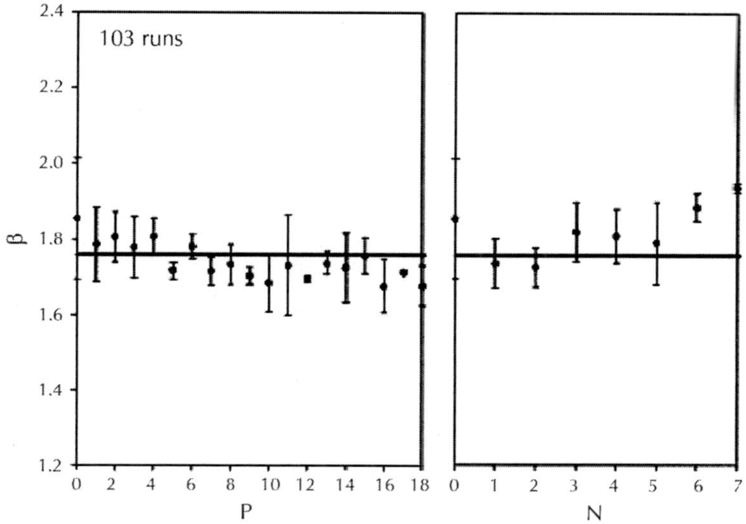

Figure 4: Parameter as a function of the number of passes (P) and number of channels per pass (N).

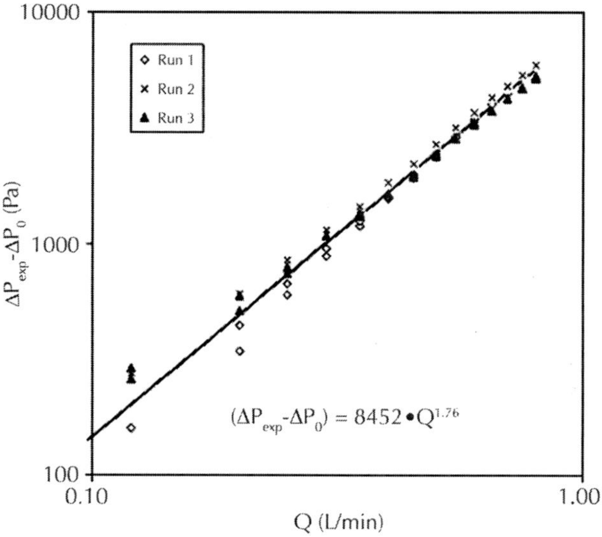

Figure 5: Example of determination of parameter α for flow arrangement 11×1.

Finally, the correlation between pressure drop and flow rate for each flow arrangement was represented by Eq. (4) (obtained from Eqs. (2), (3a) and (3b)), using parameter α'. Note that for flow arrangement 0×0, $\alpha'=0$ and, consequently, PHE = Δ0.0 Pa:

$$\Delta P_{PHE}=(\alpha-\alpha_{pipe})Q^{1.76}=\alpha'\cdot Q^{1.76} \qquad (4)$$

Correlation for the Effect of *P* and *N*

The pressure drop data of all flow arrangements could be well fitted using Eq. (4), where parameter α'depends on *P* and *N*. Fig. 6 and Table 2 shows a clear linear dependence between α' and *P* for a given number of channels per pass, *N*.

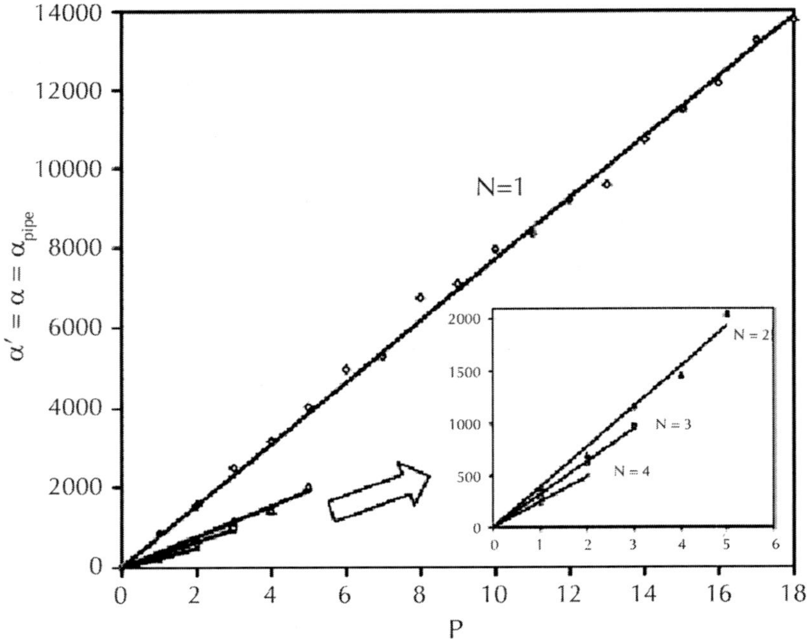

Figure 6: Influence of number of passes (P) on parameter α' for N=1, 2, 3, and 4.

Table 2: Influence of the number of passes (P) on parameter α' for given a given number of channels per pass (N=1, 2, 3, and 4)

N	Correlation	r^2
1	$\alpha'=771\cdot P$	0.997
2	$\alpha'=381\cdot P$	0.991
3	$\alpha'=315\cdot P$	0.996
4	$\alpha'=241\cdot P$	0.991

This linear dependence was expected because an arrangement P×N can be assumed equivalent to a series association of *P* arrangements 1×N, from a pressure drop point of view. Consequently, $\Delta P_{PHE}(P\times N)\approx P\cdot\Delta P_{PHE}(1\times N)$. The influence of the number of passes on the pressure drop can then be expressed as in Eq. (5). Note that the flow rate was replaced by the mean channel flow rate$_{Qc}$=Q/N and a new model parameter $\alpha''=(\alpha'\cdot N^{1.76}/P)$ is introduced:

$$\Delta P_{\text{PHE}} = \alpha' \cdot Q^{1.76} = \left(\frac{\alpha' \cdot N^{1.76}}{P}\right) P \cdot Q_c^{1.76} = \alpha'' \cdot P \cdot Q_c^{1.76} \tag{5}$$

The next step was to determine how the number of channels per pass affects parameter α''. Fig. 7 shows that this dependence can be considered linear for N ≤4; however, for N≤7 it can be well fitted by a second order polynomial. This behavior modification after N=4 can be explained by the problem of flow maldistribution, where the flow is unequally distributed between the channels of a pass resulting on higher pressure drop and lower thermal effectiveness (Bassiouny and Martin, 1984 and Rao and Das, 2004).Bobbili et al. (2006) and Tereda et al. (2007) verified that as the number of channels per pass increases, the effects of flow maldistribution are more pronounced, in accordance with the results in Fig. 7.

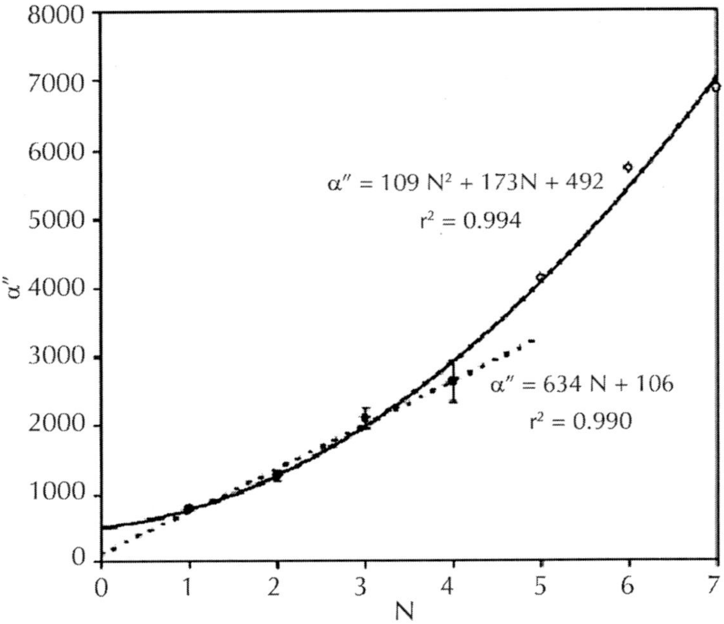

Figure 7: Influence of the number of channels per pass (N) on parameter α″.

Finally, the effect of the flow arrangement on the pressure drop of the studied PHE can be summarized by Eq. (6). This correlation between ΔP_{PHE} (Pa) and Q_c (L/min) contains four empirical parameters and shows a good fit between experimental and predicted values of pressure drop, as can be seen in the parity chart in Fig. 8. The mean absolute error for predicted pressure drop is 9.1%.

$$\Delta P_{\text{PHE}} = (109 \cdot N^2 + 173 \cdot N + 492)P \cdot Q_c^{1.76} \tag{6}$$

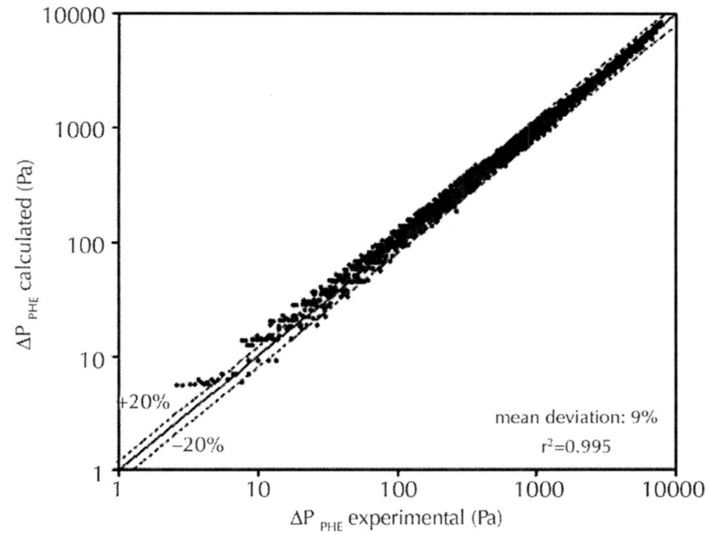

Figure 8: Parity chart of ΔP_{PHE} for the obtained correlation.

Friction Factor Model Adjustment

In addition to obtaining the empirical correlation in Eq. (6), the set of 1918 experimental points were used for adjusting the pressure drop model in Eq. (1). At first, a general correlation between the Fanning friction factor and the Reynolds number was assumed, as in Eq. (7), where *a*, *b*, and *c* are the model parameters. The Reynolds number is defined as $Re = _{De} \cdot _{vc} \cdot \rho/\mu$, and its range for the experimental data was $10 \leq Re \leq 870$, with $0.0032 \leq v_c \leq 0.26$m/s.

$$f = a + \frac{b}{Re^{c}} \tag{7}$$

The mean squared error between experimental and predicted log-values of pressure drop was minimized by varying parameters *a*, *b*, *c* and λ. Constraints for $f_i \geq 0$ (i=1, 2,..., 1918) were included in the optimization problem, which was solved by a quasi-Newton search algorithm.

Typical forms of Eq. (7) were tested, which consider a=0 (Shah and Focke, 1988 and Saunders, 1988) and/or c=1 (Edwards et al., 1974 and Leuliet et al., 1987). The form proposed by Leuliet et al. (1990), where *a* is replaced by *a/Re* in Eq. (7), was also tested. Moreover, the use of parameter λ was tested.

The criteria used for the evaluation of the model fit were: number of parameters, parity chart, mean squared error, and error distribution. Two models presented a fairly good agreement between experimental and predicted values of pressure drop. The friction factor correlation for the first adjusted model is presented in Eq. (8), with optimal parameter for duct pressure drop λ=1.44 (see Eq. (1)). This type of friction factor correlation was originally proposed for PHEs by Edwards et al. (1974) based on an analogy to pressure drop in packed beds. The friction factor correlation for the second best fit is a typical turbulent flow correlation, as shown in Eq. (9), also with duct pressure drop parameter λ=1.44:

$$f = 0.30 + \frac{29}{Re} \quad (8)$$

$$f = \frac{3.6}{Re^{0.38}} \quad (9)$$

The parity charts and mean absolute error for both fitted models are presented in Fig. 9. The distribution of experimental points in Fig. 8 and Fig. 9 is similar with a fairly good fit for high pressure drop and a larger dispersion for low pressure drop (low Reynolds flow). Approximately 15% of the experimental points have pressure drop as small as 100 Pa and thus are more affected by experimental error. However, this larger error for low Reynolds have a small influence for PHE design problems because the maximum allowable pressure drop is the limiting constraint. In addition, Fig. 10 presents the normal probability error distribution (Statistica 7.1, StatSoft, Tulsa, USA) for the prediction of the PHE pressure drop using the empirical approach in Eq. (6) and the friction factor approach in Eq. (1) using *f*—correlations in Eqs. (8) and (9). The distribution which is closer to the normal distribution is that from Eq. (9).

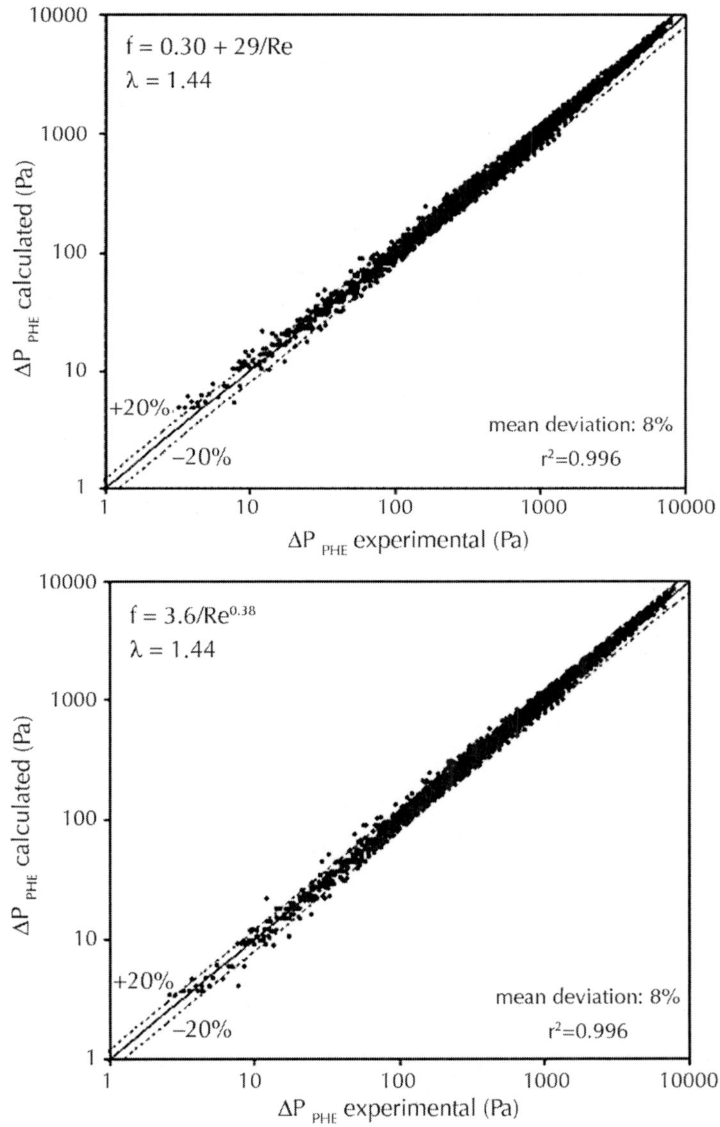

Figure 9: Parity chart of ΔP_{PHE} for the best two adjusted models.

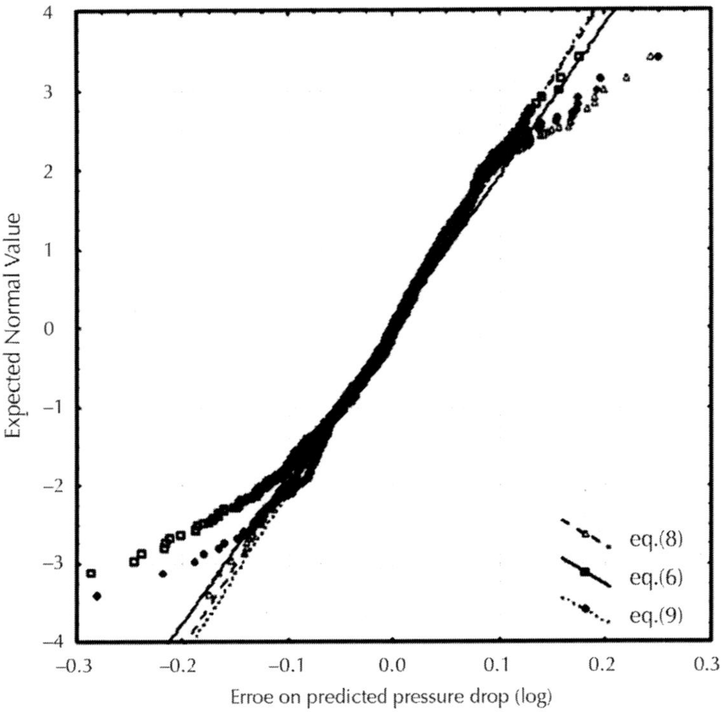

Figure 10: Normal probability distribution of error for the three adjusted equations for pressure drop prediction, $\log(\Delta P_{PHE})$.

Successfully, the obtained parameter for duct pressure drop =1.44 is in accordance with the values proposed in literature, thus validating the use of the model in Eq. (1) for evaluating the pressure drop in PHEs. With the obtained value of λ, Eq. (1) can be applied for separating the friction pressure drop from the duct pressure drop, thus allowing the determination of the experimental friction factor for each one of the 1918 experimental points. The comparison between experimental *f* and the adjusted *f*-correlations is shown in Fig. 11. The theoretical correlation $f=24/Re$ for infinite parallel flat plates is also included (Muley et al., 1999). A larger dispersion of the points can be noticed for low Reynolds, but it does not affect the quality of the pressure drop adjustments, as seen in Fig. 9.

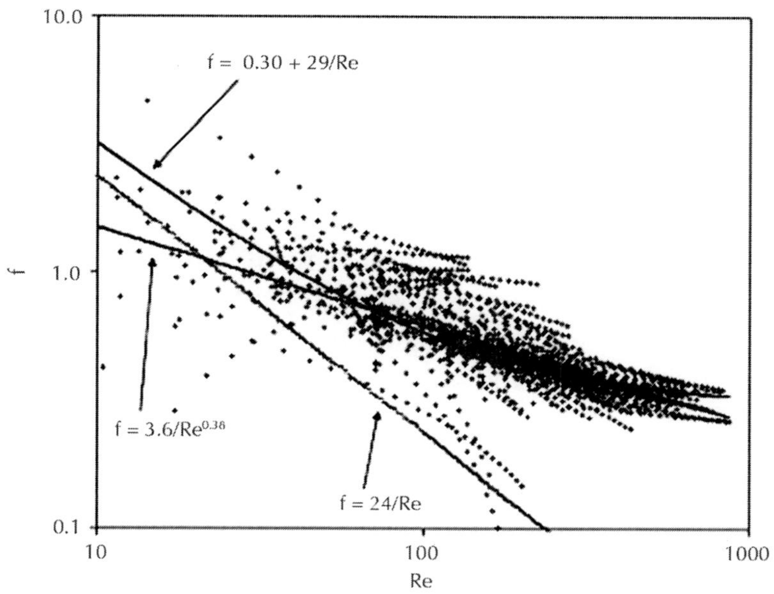

Figure 11: Adjusted models and experimental values of friction factor (f) as a function of Reynolds number (*Re*).

CFD Modeling

Galeazzo et al. (2006) performed CFD simulations of the FT-43 PHE for configurations 1×2/1×2 and2×1/2×1. The results obtained from steady state simulations of the heat transfer, in the form of outlet temperatures and heat load, were in good agreement with the experimental data for a broad range of Reynolds number $93 \leq Re \leq 894$. Moreover, temperature and velocity distributions inside the flow channels are presented by the authors. The same approach is used in the present study for the simulation of isothermal flow in one of the PHE sides for determining ΔP_{PHE} and comparing the results with experimental data.

The computational grids were generated with the software GAMBIT (Fluent Inc., Lebanon, USA) and the problem was numerically solved using the finite volume method with the software FLUENT versions 6.3 (Fluent Inc., Lebanon, USA). The fluid domains were modeled with the properties of water at 23 °C (ρ=998 kg/m^3, μ=9.36 × 10^{-4} Pas). The boundary condition at the inlet was the volumetric flow rate and at the outlet was atmospheric pressure. Although the flow is mostly laminar inside the PHE, there are turbulent regions near the plate ports and inside the interconnecting channels, making the use of turbulence models mandatory. The performance of the k–ε and the shear stress transport (SST) turbulence models was assessed. Both turbulence models were used in their low-Reynolds number formulations, because the small distance between the plates makes the fulfillment of the y^+ criteria impossible for the wall function approach. Second order discretization of all variables was employed, and the simulation was considered converged when the residual of all variables remained below 10^{-4} and the mean outlet velocities remained constant.

An exploded view of the computational domain can be seen in Fig. 12. The grid sensitivity of the results was checked comparing the results of two grids with different refinement levels using the k–ε turbulence model. The coarse grids for the 1×2/1×2 and 2×1/2×1 configurations had approximately 0.6 million hexahedral elements, while the fine grids had approximately 1.6 million hexahedral elements. The mean absolute difference between the results with the course and fine grids was 1.7% for the 1×2/1×2configuration and 6.4% for the 2×1/2×1 configuration. As the results showed small but significant grid sensitivity, the fine grid was used in the rest of the work.

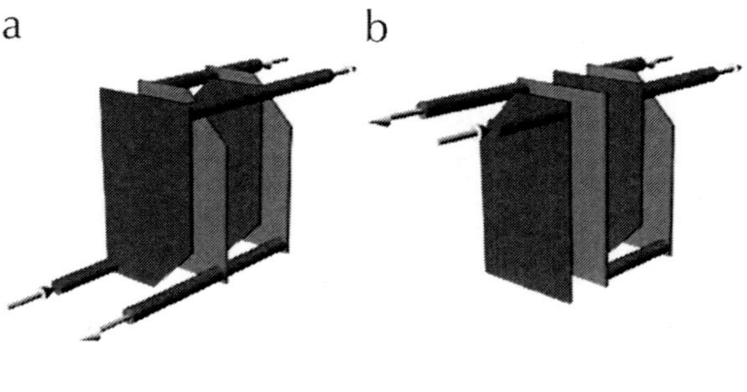

Figure 12: Exploded view of the CFD domain for configurations (a) 1×2/1×2 and (b) 2×1/2×1 (metal plates are not shown).

The comparison of the results obtained with the k–ε and the SST turbulence models showed that the SST model predicts the pressure drop being in average 5% greater than the pressure drop predicted using the k–ε turbulence model. This difference can be attributed to the ability of the SST turbulence model to better resolve the transition between laminar and turbulent flows.

The results are presented in Fig. 13, where simulations using the fine grid and the SST turbulence model were compared to the measurements using 10 different inlet flow rates in the range 0.10–1.45 L/min. A good agreement between experimental and predicted values of pressure drop was obtained. However, it can be seen in Fig. 13 that the predicted pressure drop is close to the experimental runs with lower pressure drop. As discussed before, the measured pressure drop is very sensitive to the slight misalignment of gaskets, which increases the pressure drop. Since the virtual PHE does not have this problem, the CFD simulations under predict the pressure drop, as can be seen in Fig. 13. Nevertheless, the mean deviations (13% for arrangement 1×2 and 10% for 2×1) does not compromise the quality of the CFD results.

The CFD model of arrangement 1×2 was further applied for evaluating the distribution of flow between the first and the second channels of the pass. For the design of PHEs it is usually assumed that the flow is equally distributed through the channels of a pass; however, flow maldistribution is common in practice and reduces the PHE performance (Bobbili et al., 2006 and Tereda et al., 2007). Fig. 14 shows the relation between the flow rate on the first channel and the inlet flow rate (Q_1/Q) as a function of Q. Surprisingly, the maldistribution is significant even for the PHE with only two channels per pass. For the maximum flow rate (Q = 1.45 L/min), the first channel receives 59% of the inlet flow rate. PHEs with a larger number of channels should be simulated using CFD tools for better evaluating maldistribution and its effect on the performance of the exchanger; nevertheless, the complexity of the virtual model of the PHE is constrained by the current computational power.

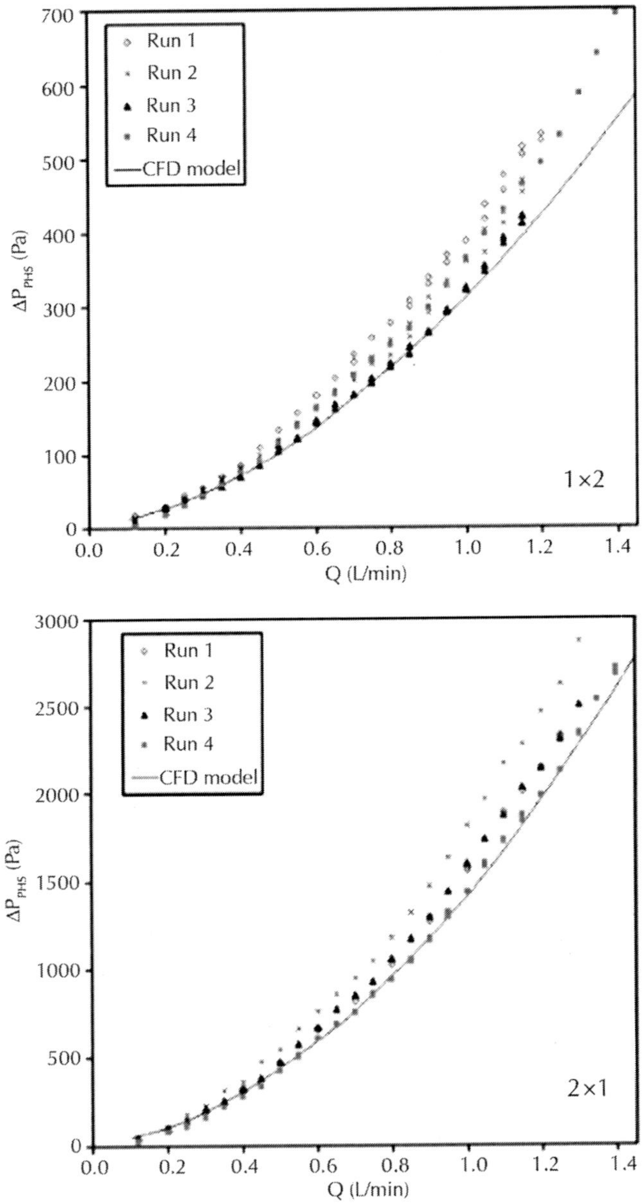

Figure 13: Experimental data and CFD predictions of pressure drop for flow arrangements 1×2 and 2×1.

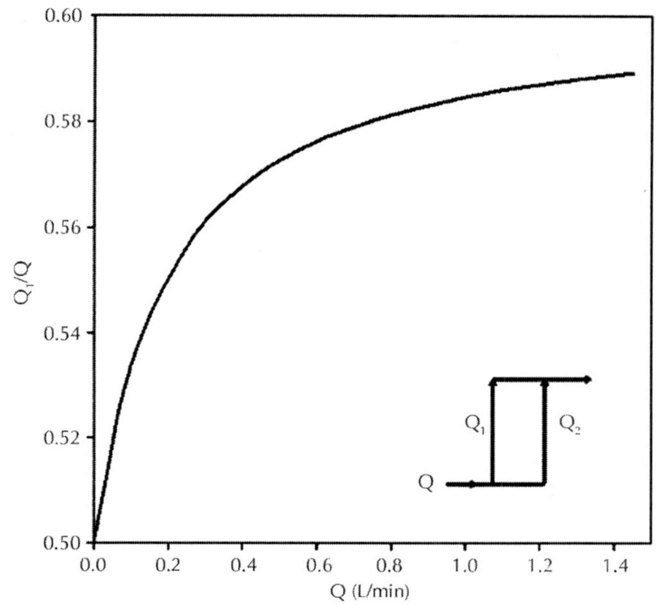

Figure 14: Distribution of flow for arrangement 1×2, where Q_1 is the flow rate on the first channel.

CONCLUSIONS

Isothermal pressure drop of 32 different flow arrangements were experimentally obtained using a laboratory scale PHE with flat plates without corrugation. It was possible to correlate the pressure drop with the flow rate Q, number of passes P, and number of flow channels per pass N (see Eq. (6)). A linear dependence between P and the pressure drop was verified, as expected. However, the effect of N was not clear, possibly due to flow maldistribution among flow channels. The experimental data was also used for adjusting and validating a friction factor type model for pressure drop. For this model it is assumed that the pressure drop within the plate ports is proportional to the inlet velocity head. The adjusted parameter for this proportionality is very close to the values suggested in literature. Moreover, the experimental data of pressure drop for arrangements

1×2 and 2×1 were compared with the predictions of a CFD model of the PHE, where also a good agreement between experimental and predicted values was obtained. The CFD model was also applied for evaluating the flow maldistribution for arrangement 1×2.

ACKNOWLEDGMENTS

The authors would like to acknowledge financial support from FAPESP (The State of São Paulo Research Foundation).

REFERENCES

1. Bassiouny, M.K., Martin, H., 1984. Flow distribution and pressure drop in plate heat exchangers—I U-type arrangement. Chemical Engineering Science 39 (4), 693–700.
2. Bobbili, P.R., Sunden, B., Das, S.K., 2006. An experimental investigation of the port flow maldistribution in small and large plate package heat exchangers. Applied Thermal Engineering 26 (16), 1919–1926.
3. Edwards, M.F., Changal Vaie, A.A., Parrot, D.L., 1974. Heat transfer and pressure drop characteristics of a plate heat exchanger using Newtonian and non-Newtonian liquids. Chemical Engineer 285, 286–293.
4. Galeazzo, F.C.C., Miura, R.Y., Gut, J.A.W., Tadini, C.C., 2006. Experimental and numerical heat transfer in a plate heat exchanger. Chemical Engineering Science 61 (21), 7133–7138.
5. Gut, J.A.W., Pinto, J.M., 2004. Optimal configuration design for plate heat exchangers. International Journal of Heat and Mass Transfer 47 (22), 4833–4848.
6. Gut, J.A.W., Fernandes, R., Pinto, J.M., Tadini, C.C., 2004. Thermal model validation of plate heat exchangers with generalized configurations. Chemical Engineering Science 59 (21), 4591–4600.

7. Kakaç, S., Liu, H., 2002. Heat Exchangers: Selection Rating and Thermal Design. second ed. CRC Press, Boca Raton. pp. 323–354.
8. Leuliet, J.C., Maingonnat, J.F., Lalande, M., 1987. Etude de la perte de charge dans des échangeurs de chaleur à plaques traitant des produits non-Newtoniens. Revue Générale de Thermique 26 (308–309), 445–450.
9. Leuliet, J.C., Maingonnat, J.F., Lalande, M., 1990. Écoulements et transferts de chaleur dans les échangeurs à plaques traitant des produits visqueux Newtoniens et pseudoplastiques, 1. modélisation des variations du diamètre hydraulique. Canadian Journal of Chemical Engineering 68 (2), 220–229.
10. Muley, A., Manglik, R.M., Metwally, H.M., 1999. Enhanced heat transfer characteristics of viscous liquid flows in a chevron plate heat exchanger. Journal of Heat Transfer 121 (4), 1011–1017.
11. Rao, B.P., Das, S.K., 2004. Effect of flow distribution to the channels on the thermal performance of the multipass plate heat exchangers. Heat Transfer Engineering 25 (8), 48–59.
12. Saunders, E.A.D., 1988. Heat Exchangers Selection, Design & Construction. Harlow, Longman Scientific & Technical, pp. 104–116, 364–373.
13. Shah, R.K., Focke, W.W., 1988. Plate heat exchangers and their design theory. In: Shah, R.K., Subbarao, E.C., Mashelkar, R.A. (Eds.), Heat Transfer Equipment Design. Hemisphere P.C., pp. 227–254.
14. Tereda, F.A., Srihari, N., Sunden, B., Das, S.K., 2007. Experimental investigation on port-to-channel flow maldistribution in plate heat exchangers. Heat Transfer Engineering 28 (5), 435–443.

Chapter 9

Design of Multiple Shell and Tube Heat Exchangers in Series: E Shell and F Shell

U. Vengateson

Lurgi India Co. Pvt. Ltd., A-24/10 Mohan Cooperative Industrial Estate, Mathura Road, New Delhi 110044, India.

ABSTRACT

Multiple shell and tube heat exchangers in the series are employed to handle the temperature cross in the chemical process industries. Depending on the degree of temperature cross, certain number of heat exchangers (either E or F shell type) need to be connected in series such that the temperature cross in each exchanger is within allowable limit. Determination of the number of exchangers for the given terminal temperatures is essential during heat exchanger design phase. In this paper, using finite difference calculus, modeling has been done to calculate the number of shells required for both E and F shell cases. In addition, equations are developed to determine hot and cold fluid temperature profiles across all heat exchangers. Design procedure is illustrated with the help of a case study and the capital cost of both cases is compared. Issues related to E shell and F shells are also discussed.

INTRODUCTION

There are several shell configurations designated as E, F, G, H, J, K and X by the Tubular Exchanger Manufactures' Association Inc. These are described in detail in literature (Perry and Green, 1997). E shell is a single-pass shell, and the number of tube passes may be one or multiples of two (two is most common). The shell side fluid enters at one end and leaves the other end of the opposite side. F shell is a two-pass shell that has a longitudinal baffle dividing the shell into two compartments, shell fluid enters at one compartment, travels the entire length of the shell through that compartment, turns around and flows through the another compartment of the shell and finally leaves at the same end of the other side. The number of tube passes for F shell may be two or multiples of four (four is most common). Considering first the 1–2 heat exchanger in Fig. 1(a), the tube fluid in the first tube pass is in parallel with the shell fluid, and in the second tube pass the tube fluid is in the counter flow with the shell fluid. Hence, the log mean temperature difference (LMTD), which applies to either parallel or counter flow but not to a mixture of both types, cannot be used to calculate the true or effective mean temperature difference (EMTD) without correction. Similarly for 2–4 heat exchanger, as may be seen in Fig. 1(b), contact between shell fluid and tube fluid is a mixture of both parallel and counter flows and hence correction factor is necessary to get the EMTD. This EMTD is generally obtained by multiplying

LMTD of true counter current flow by a flow correction factor (F_T). This factor is correlated in terms of two dimensionless ratios, R and P by Nagle (1933) and Underwood (1934). For instance, derivations can be found in Kern (1997) and following assumptions were made during the derivation: stream flows are at steady state, overall heat transfer coefficient and specific heat remain constant throughout the exchanger, there is no phase change and heat losses are negligible. Eqs. (1) and (2),Kern (1997) are used to calculate F_T for 1–2 exchanger and F_T for 2–4 exchanger, respectively.

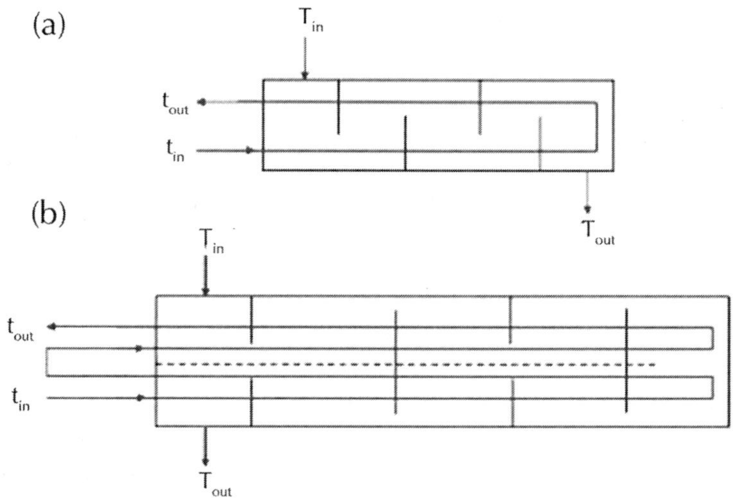

Figure 1: Schematic diagram of E shell and F shell: (a) 1–2 heat exchanger; (b) 2–4 heat exchanger.

For 1–2 heat exchanger

$$= \frac{\sqrt{1+R^2}\ln[(1-P)/(1-PR)]}{(R-1)\ln[(2-P(R+1-\sqrt{1+R^2}))/(2-P(R+1+\sqrt{1+R^2}))]} \tag{1}$$

$$\text{for } R = 1 \quad F_T = \frac{\sqrt{2}P}{(1-P)\ln[2-P(2-\sqrt{2}))/(2-P(2+\sqrt{2})]} \tag{1a}$$

For 2–4 heat exchanger

$$= \frac{[\sqrt{1+R^2}/2(R-1)]\ln[(1-P)/(1-PR)]}{\ln[2/P-1-R+(2/P)\sqrt{(1-P)(1-PR)} + \sqrt{1+R^2})/(2/P-1-R+(2/P)\sqrt{(1-P)(1-PR)}-\sqrt{1+R^2}]} \quad (2)$$

$$\text{for } R = 1 \quad F_T = \frac{P}{\sqrt{2}(1-P)\ln[(4-4P+P\sqrt{2})/(4-4P-P\sqrt{2})]} \quad (2a)$$

R is the ratio of the thermal capacity of cold and hot fluid

$$R = \frac{mc_p}{MC_p} = \frac{T_{in} - T_{out}}{t_{out} - t_{in}} \quad (3)$$

P is the thermal effectiveness of the exchanger and will be always less than 1.

$$P = \frac{t_{out} - t_{in}}{T_{in} - t_{in}} \quad (4)$$

F_T can be found for a given value of R and P from above correlations and these values are plotted in Fig. 2 and Fig. 3 for 1–2 and 2–4 exchangers, respectively. Values of F_T are not significantly decreased further by using exchangers with additional tube passes. For example F_T for 1–8 exchanger differs by less than 2% from that for a 1–2 exchanger (Seider et al., 2004). So F_T of E shell (1–2, 1–4, 1–6, 1–8 exchangers) is almost same. Similarly F_T of F shell (2–4, 2–8 exchangers) can be considered as same.

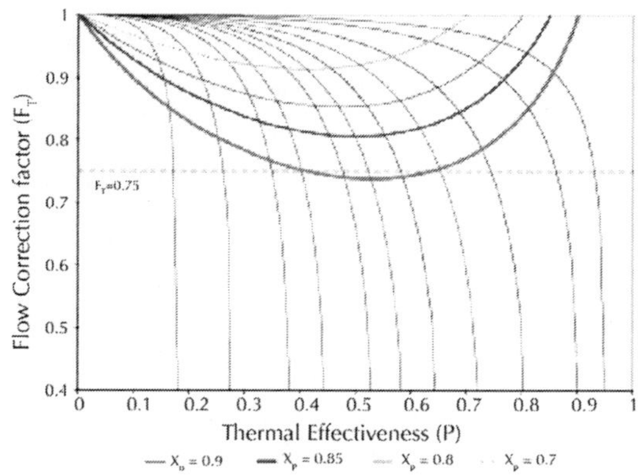

Figure 2: Locus of constant X_P value on F_T correction factor chart for 1–2 shell and tube heat exchanger.

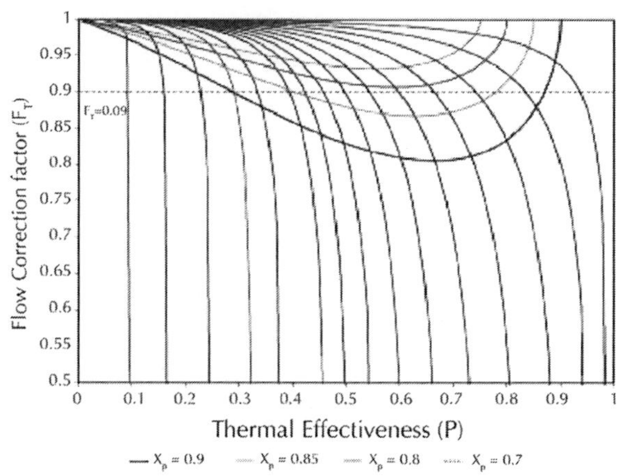

Figure 3: Locus of constant X_P value on F_T correction factor chart for 2–4 shell and tube heat exchanger.

Temperature Approach and Temperature Cross

Based on terminal temperatures of hot and cold fluids, there are two situations in heat exchanger design—temperature approach and temperature cross. If the final temperature of the cold stream is lower than the hot fluid outlet temperature for counter current flow as shown in Fig. 4(a), then it is the temperature approach. On the other hand, if final temperature of the cold stream is higher than the hot fluid outlet temperature for counter current flow as shown in Fig. 4(b), then it is treated as temperature cross (Smith, 2005). Obviously, there can never be temperature cross for co-current flow since it violates second law of thermodynamics. Heat exchanger design for temperature approach situation is straightforward since it can always be accommodated. Problem arises only in the temperature cross depends on whether it is small or large. Wales (1981) proposed a new parameter G which is defined as the ratio of outlet temperature gap to the inlet temperature difference.

$$G = \frac{T_{out} - t_{out}}{T_{in} - t_{in}} \tag{5}$$

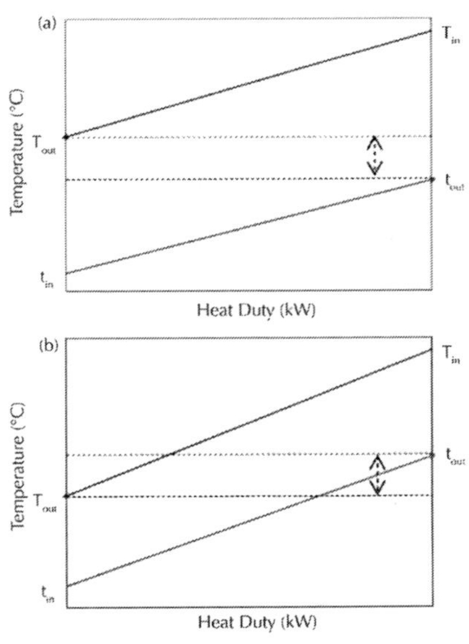

Figure 4: Temperature approach and temperature cross situations in heat exchanger design: (a) temperature approach ($t_{out} < T_{out}$); (b) temperature cross ($t_{out} > T_{out}$).

This parameter can be related to parameters R and P in Eq. (6).

$$G = 1 - P(1 + R) \tag{6}$$

Values of G can range from +1 (no heat exchange) to −1 (highest heat exchange). Positive G value represents temperature approach and negative G value represents temperature cross. G becomes 0 when outlet temperatures of both fluids are same, i.e., temperature meet. F_T can be related through G and R using Eqs. (1), (1a), (2), (2a), (3), (4), (5) and (6) and the graphical representations for E shell and F shell exchangers are shown in Fig. 5 and Fig. 6 respectively. It is clear from the Fig. 5 and Fig. 6 that F_T decreases moderately with decreasing positive G values, but falls sharply both where the temperature meet (G = 0) and where the G values are negative (temperature cross). It is also evident that the higher the R, the sharper is the fall in F_T. The only advantage of this method is G itself provides more information about the value of F_T rather than does either P or R alone.

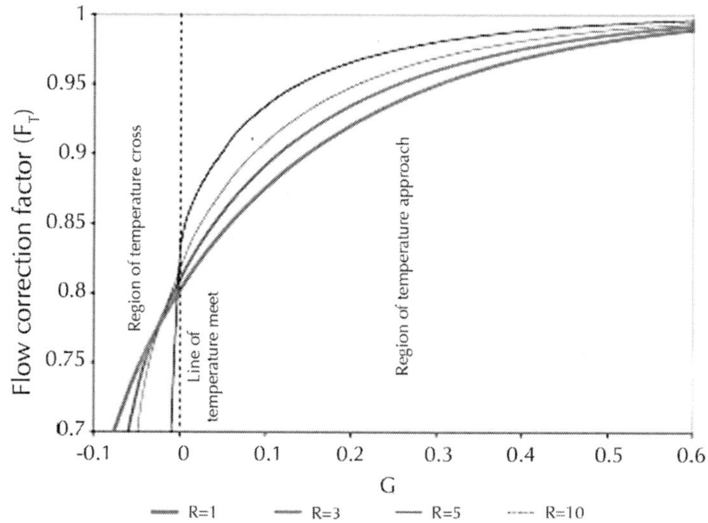

Figure 5: F_T correction factor chart for E shell based on R and G.

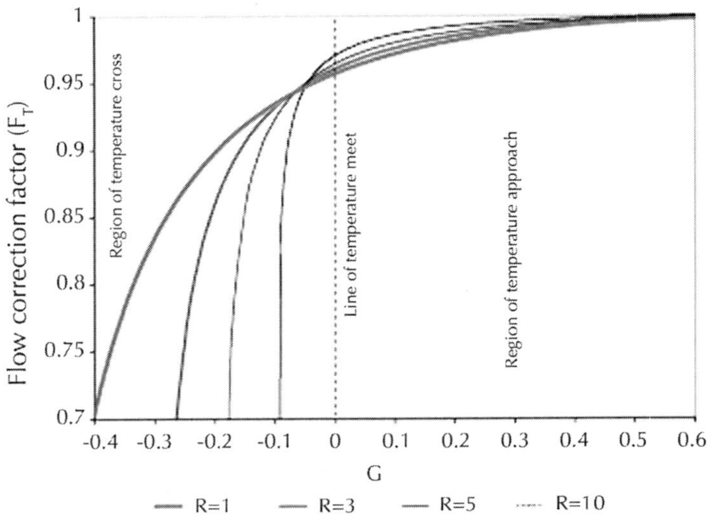

Figure 6: F_T correction factor chart for F shell based on R and G.

As the amount of temperature cross increases, the F_T decreases significantly, causing a dramatic increases in the heat transfer area. The increase in heat transfer area is severe when F_T is less than 0.75 for 1–2 heat exchanger as can be seen from that F_T curve slope becomes very steep. The low value of F_T indicates inefficient use of the heat transfer area. It can be avoided using multiple shell and tube heat exchangers connected in series. Single F shell, of course can handle larger temperature cross than E shell since its $F_T$value is greater than F_T value of E shell for same R and P. For example consider a case where hot fluid is cooled from 80 °C to 45 °C and cold fluid is heated from 30 °C to 47 °C. Here R = 2.05, P = 0.34. F_T values are 0.76 for E shell and 0.95 for F shell. If temperature cross is too large, single F shell itself inefficient. Hence, N number of either E shells or F shells needs to be connected in series so that F_T value of each shell is more than critical value. The desired values are $F_T \geq$ (0.75 or 0.8) for E shell and $F_T \geq 0.9$ for F shell since these values stay away from the steep slope region. The number N depends on degree of temperature cross. Theoretically, N tends to ∞ when cold fluid outlet temperature tends to hot fluid inlet temperature. As can be seen from Fig. 2 and Fig. 3, for each value of

R, there is a maximum possible value of P (P_{max}). Similarly for each value of R, there is a minimum possible value of G (G_{min}) as shown in Fig. 5 and Fig. 6. Mathematically, value of F_T when P tends to 0 is 1, for any R and this value tends to $-\infty$ when P is increased to P_{max} for the same value of R. In other words, increase of P from 0 to P_{max} (or decrease of F_T from 1 to $-\infty$) for fixed Rcauses increase of heat transfer area from a minimum value to ∞. F_T tends to $-\infty$ is just a representation thatF_T curve falls vertically downwards, in fact $F_T < 0$ has no physical meaning. For large temperature cross situations, the value of P may exceed P_{max}. It is essential to fix P of individual exchanger ($P < P_{max}$) so that it causes F_T value to exceed critical value. Lower value of individual exchanger P indicates more number of exchangers with lesser area is required to be connected in series.

Selection of P of Individual Exchanger

Ahmad et al. (1988) introduced a parameter X_P, to fix the P value of each shell by multiplying it with P_{max} for a given value of R, i.e., ($P = X_P \cdot P_{max}$). Ahmad et al. (1988) also presented another method to fix the P. In this method, for a given value of R, P is selected so that slope of F_T curve at (P, F_T) becomes −2.8. The value −2.8 is basically slope of F_T curve at $F_T = 0.75$, for R = 1. Though this method provides guarantee to stay away from the steep region in the F_T curve, the analytical solution to solve P by equating $(\partial F_T/\partial P)$ R with −2.8 is very complex. In order to obtain P value readily, Shenoy (1995) developed Eq. (7) by curve fitting the data obtained from the constant slope criterion. This expression maintains the slope close to −2.8, though the actual slope obtained from this expression varies between −2.75 and −2.93.

$$X_{PP} = 1 - \frac{0.223}{\left[1 + (0.223/(0.033 + 0.103R))^{1.4}\right]^{1/1.4}} \tag{7}$$

He also pointed out constant X_P curve and constant slope $(\partial F_T/\partial P)R = -2.8$ curve do not have identical profile and for some cases, design accepted in one approach are rejected in another approach. He developed another approach based on constant slope in F_T ($X_{P,}$R) chart rather than conventional F_T (PR) chart. In this approach locus of constant slope $(\partial F_T/\partial X_P)R = -1.64$ is drawn in F_T (P, R) chart and this curve is explicitly expressed in terms of R in Eq. (8). The value −1.64 is the slope of F_T curve atF_T = 0.75 and R = 1 in F_T ($X_{P,}$R) chart.

$$X_{PC} = 1 - 0.1\exp[-0.5(\log R)^2] \tag{8}$$

All these parameters – $X_{P,}X_{PP,}X_{PC}$ – proposed by Ahmad et al. (1988) and Shenoy (1995) are basically fractional multiplier of P_{max} to get P of individual exchanger and these are not directly linked to temperature cross. Comparisons of these approaches are shown in Fig. 7. Required number of shells based on these three approaches is sometimes different and subsequently affects total heat transfer area and heat exchanger cost. In order to minimize the total cost, Moita et al. (2004) introduced a design algorithm, to allow the best choice among these three X_P approaches. Gulyani (2000) presented a new approach to choose Pthough the parameter G defined in Eq. (5). In this approach, Gulyani (2000) introduced a parameter Y that directly accounts temperature cross and the required G for individual exchanger can be obtained in Eq. (9).

$$G = G_{min} + Y \tag{9}$$

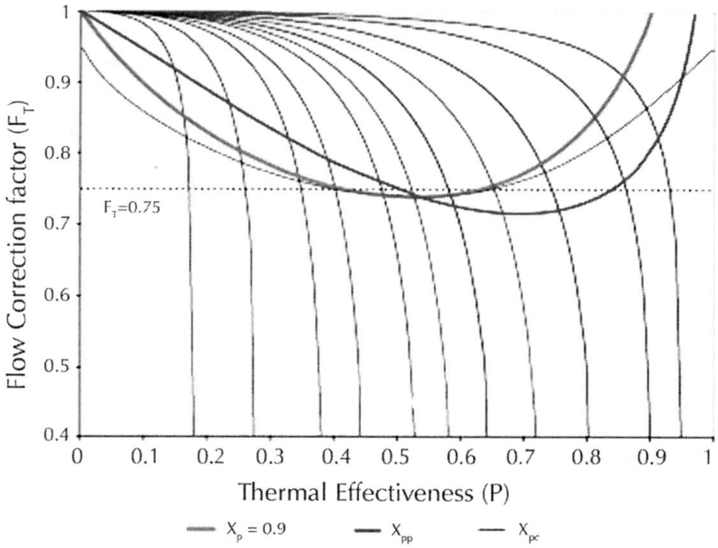

Figure 7: Comparison of different approaches – constant X_P (X_P = 0.9), $X_{PP,}X_{PC}$ – to select individual exchanger P and F_T.

G_{min} represents the maximum temperature cross theoretically feasible in E shell, for a given value of R. Obviously this G_{min} can be obtained through Eq. (6), when $P = P_{max}$. The parameter Y is the constant set by the designer. If the designer wants zero temperature cross, i.e., temperature meet, then Y becomes $-G_{min}$ so that G = 0. To allow some temperature cross in E shell, a value of Y in the range of 0.1–0.15 may be selected which is compatible with other design practices ($F_T > 0.75$; $X_P = 0.9$). The parameter 'Y' can be expressed in terms of X_P and R by the Eq. (10). In Gulyani (2000) approach also, individual exchanger P is obtained by multiplying X_P with P_{max}, but the only difference here is $X_P = f$ (Y, R) as can be seen from Eq(10).

$$Y = \frac{2(1+R)(1-X_P)}{(1+R)+\sqrt{1+R^2}} \quad (10)$$

Pros and Cons of F Shell

If F shell is used instead of E shell, required number of exchangers can be reduced. In addition to handling larger temperature cross than E shell, F shell is desired when shell side flow rate is low. When E shell is used for relatively low shell side flow rate, shell side velocity also becomes low, even with the smallest baffle spacing. In such situations, the allowable shell side pressure drop cannot be utilized properly. The shell side heat transfer coefficient and thus overall heat transfer coefficient becomes low, resulting in large and expensive heat exchanger. Additionally, if the shell side stream is dirty, low shell side velocity will result in heavy shell side fouling, which would require high operating costs. Since F shell has two shell passes, area for flow is less for given shell side flow rate and thus increases shell side velocity.

Though F shell is suitable for temperature cross and low shell side flow rate, there are two inherent problems associated with F shell—thermal leakage and physical (fluid) leakage. As the longitudinal baffle is exposed to the hot end of the shell stream one side and to the cold end of the shell stream on the other, a thermal conduction heat exchange decreases the effectiveness of the temperature profile. A method to account for this inefficiency is described by Rozenman and Taborek (1971). Physical leakage is due to spacing between longitudinal baffle and shell wall. It can be avoided by welding the longitudinal baffle with shell wall. However, designers commonly use flexible strips if the tube bundle must be removed for cleaning. These seals are often damaged during tube bundle removal and consequently leakage deteriorates temperature profile.

Design of Multiple Shell and Tube Heat Exchangers

The designs of multiple shell and tube exchangers involve the determination of number of shells required and the selection of heat exchanger geometry (shell diameter, tube diameter, tube length, tube layout, baffle spacing, baffle cut, etc.) which provide required heat transfer area for specified duty. A typical solution for this problem involves trial and error graphical method to determine number of shells connected in series so that F_T value of each exchanger (E shell) is equal or greater than 0.75. Then, area is found from the basic heat transfer equation if overall heat transfer coefficient is known. Some designers determine number of E shells required, by assuming equal outlet temperatures, i.e., zero temperature cross in each shell. For example, Mukherjee (1998) used this method to handle substantial temperature cross (140 °C) and arrived at four E shell which is shown schematically in Fig. 8. In this method, number of shells is estimated by counting staircases constructed alternating between heat and cold release curves. Ahmad et al. (1988) gave an analytical expression Eq. (25) to calculate number of E shells to be connected in series, for given boundary temperatures. Ponce-Ortega et al. (2008) used this solution to develop an economic optimization algorithm for E shell.

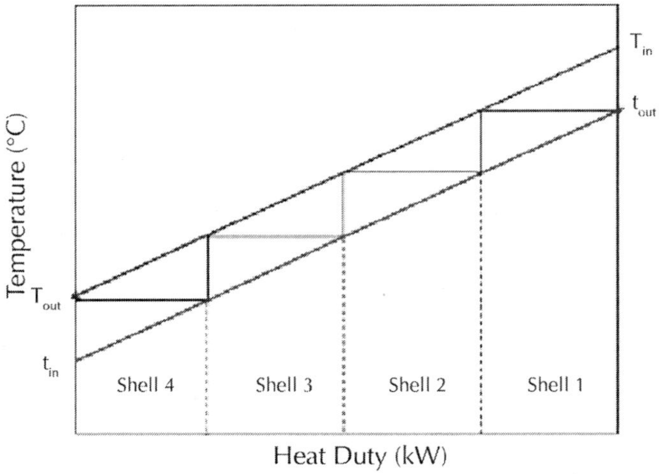

Figure 8: Traditional graphical method to determine number of 1–2 heat exchangers (zero temperature cross, i.e., temperature meet is assumed in all shells).

Though P_{max} and hence required number of exchangers (E shell) are discussed widely in many sources, for example (Ahmad et al., 1988, Gulyani, 2000, Moita et al., 2004, Smith, 2005 and Ponce-Ortega et al., 2008), issues related to maximum P value (P_{max}) of F shell and hence required number of F shells, are not addressed in open literature. In this paper, finite difference method is used to derive the equation to calculate number of shells required for both cases—E shells connected in series and F shell connected in series. Then, the same finite difference approach is extended to formulate equations that can be used to identify terminal temperatures of each exchanger. Temperature profiles of hot and cold fluids can be obtained using these equations. Design procedure is illustrated with the help of a case study and the capital cost of both cases is compared.

MATHEMATICAL MODEL

Determination of Maximum P, (P_{max}) and Multiplier X_P

From the Fig. 2 and Fig. 3, it is evident that for a given value of R, increase of P causes F_T to decrease. For each R, there is a maximum P (P_{max}) when F_T reaches its asymptotic value. The maximum value of thermal effectiveness (P_{max}) for given heat capacity ratio R can be calculated from Eq. (1) and (2) for E shell and F shell as P tends to P_{max}, F_T tends to $-\infty$. For F_T to be determinate, the following condition needs to be satisfied:

- $P < 1$
- $RP < 1$
- Argument of the logarithmic part in the denominator of Eqs. (1) and (2) should be positive.

The maximum value of P, (P_{max}) can be obtained by equating denominator of the logarithmic part to 0, as shown below:

For 1–2 heat exchanger

$$2 - P_{max}(R + 1 + \sqrt{1 + R^2}) = 0$$

$$P_{max} = \frac{2}{R + 1 + \sqrt{1 + R^2}} \tag{11}$$

Similarly, for 2–4 heat exchanger

$$2/P_{max} - 1 - R + (2/P_{max})\sqrt{(1 - P_{max})(1 - RP_{max})} - \sqrt{1 + R^2} = 0$$

$$P_{max} = \frac{2(1 + R)(1 + R^2) - 2\sqrt{(1 + R^2)(R^4 - 2R^3 + 3R^2 - 2R + 1)}}{4R + 4R^3 - R^2} \tag{12}$$

For 'N' number of 1–2 shells in series, R value is constant for each shell since ratio of specific heat for hot and cold fluids is assumed to be constant. This R is same as across all shells. P value is kept constant for each shell but it is not equal to overall P value. What value of P we can fix for each shell? We cannot fix more than P_{max}. F_T is too low and design is infeasible at $P = P_{max}$. When P is reduced below P_{max}, F_T starts increasing. Though P can be reduced by multiplying P_{max} with any one of the fractional multipliers (constantX_P, X_{PP}, X_{PC}, X_P through Y), constant X_P approach is used in this paper for further discussion. F_T value is generated for various R for different values of X_P. The locus of $X_P = 0.9$, $X_P = 0.85$, $X_P = 0.8$, $X_P = 0.7$ for E shell are shown in Fig. 2. From Fig. 2, it is clear that $X_P = 0.9$ guarantees F_T value more than 0.75 for any value of R and hence design is feasible. When X_P is reduced for fixed R, F_T starts increasing which requires lesser heat transfer area but more number of heat exchangers are required to be connected in series. A value of $X_P = 0.9$ is reasonable for multiple 1–2 shells design, but it gives F_T close to its conservative value 0.75, for R values from 0.6 to 2. For this range of R, we can choose $X_P = 0.875$. Similarly, Locus of $X_P = 0.9$, $X_P = 0.85$, $X_P = 0.8$, $X_P = 0.75$ for F shells are shown in Fig. 3. From this Fig. 3, we can choose appropriateX_P value to avoid steep slope of F_T curve. For example $X_P = 0.85$ for $R \leq 0.6$, $X_P = 0.8$

for (0.6 < R ≤ 2), $X_P = 0.85$ for R > 2, gives guaranteed value of $F_T \geq 0.9$ and avoids steep slope region. Obviously we cannot design F shell with $F_T = 0.8$ as E shell, since F_T is very close to its asymptotic value, as shown in Fig. 3.

Finite Difference Model for Number of Shells

To avoid large temperature cross in a single shell, let us break it into N number of exchangers, so that $F_T$value of each exchanger is more than critical value. N number of heat exchangers is connected in series as shown in Fig. 9. Hot fluid enters first exchanger and leaves N[th] exchanger. Cold fluid enters N[th] exchanger and exits 1st exchanger. Let 'n' represents any integer, 1 ≤ n ≤ N. POn and ROn represent overall effectiveness and heat capacity ratio up to 'n' heat exchangers. Similarly, $PO_{(n-1)}$ and $RO_{(n-1)}$ represent overall effectiveness and heat capacity ratio up to the (n − 1) heat exchangers. Pn and Rn are effectiveness and heat capacity ratio of nth heat exchanger.

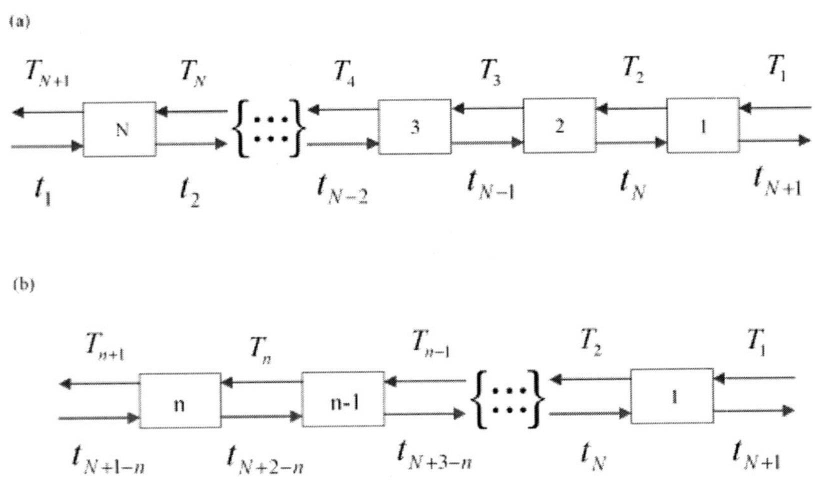

Figure 9: Heat exchangers connected in series: (a) hot and cold fluid temperatures for N number of exchangers; (b) hot and cold fluid temperatures up to 'n' exchangers; n is an integer, 1 ≤ n ≤ N.

Expressions for POn, ROn, $PO_{(n-1)}$, $RO_{(n-1)}$, Pn, Rn are given in Eq. (13), (14), (15), (16), (17) and (18)

$$P_{On} = \frac{t_{N+1} - t_{N+1-n}}{T_1 - t_{N+1-n}} \tag{13}$$

$$R_{On} = \frac{T_1 - T_{n+1}}{t_{N+1} - t_{N+1-n}} \tag{14}$$

$$P_{O(n-1)} = \frac{t_{N+1} - t_{N+2-n}}{T_1 - t_{N+2-n}} \tag{15}$$

$$R_{O(n-1)} = \frac{T_1 - T_n}{t_{N+1} - t_{N+2-n}} \tag{16}$$

$$P_n = \frac{t_{N+2-n} - t_{N+1-n}}{T_n - t_{N+1-n}} \tag{17}$$

$$R_n = \frac{T_n - T_{n+1}}{t_{N+2-n} - t_{N+1-n}} \tag{18}$$

Effectiveness of each individual exchanger is assumed to be equal but it is not equal to overall effectiveness, and from now called P

$$P = P_1 = P_2 = P_3 = \cdots = P_{n-1} = P_n = P_N \tag{19}$$

Heat capacity ratio of each individual exchanger is equal and also same as overall capacity ratio, and from now called R,

$$R = R_1 = R_2 = R_3 = \cdots = R_{n-1} = R_n = R_N = R_{On} = R_{ON} \tag{20}$$

From Eqs. (13), (14), (15), (16), (17), (18), (19) and (20), Eq. (21) is obtained.

$$\left[\frac{1-P_{On}R}{1-P_{On}}\right]-\left[\frac{1-PR}{1-P}\right]\left[\frac{1-P_{O(n-1)}R}{1-P_{O(n-1)}}\right]=0 \tag{21}$$

$$\text{Let}\, Z_n=\frac{1-P_{On}R}{1-P_{On}} \tag{22}$$

Then Eq. (21) becomes

$$Z_n-\left[\frac{1-PR}{1-P}\right]Z_{n-1}=0 \tag{23}$$

Eq. (23) is a finite difference equation of order 1. The solution contains only complementary function and no particular integral, and it is given in Eq. (24). Details of solution procedure can be found in Rice and Do (1995).

$$Z_n=\left[\frac{1-PR}{1-P}\right]^n \tag{24}$$

After substituting n = N, $P = X_P \cdot P_{max}$ and taking logarithm on both sides of Eq. (24), it becomes Eq. (25).

$$N_{SHELLS}=\frac{\ln[(1-RP_{ON})/(1-P_{ON})]}{\ln[(1-RX_PP_{max})/(1-X_PP_{max})]} \tag{25}$$

Eq. (25) can be used to calculate theoretical number of shells required for given boundary temperatures for both E shell and F shell. The only difference is $P_{max} \cdot P_{max}$ for E shell and F shell are obtained from Eq. (11) and Eq. (12) respectively. Comparison of P_{max} between E shell and F shell can be seen in Fig. 10. Similar toP_{max}, G_{min} (which can be calculated from Eq. (6), when $P = P_{max}$) also differs for both E shell and F shell, as can be seen from Fig. 11. Eq. (25a) can be used for R = 1. It is determined using L'Hospital's rule as R tends to 1.

$$N_{SHELLS} = \frac{P_{ON}(1 - X_P P_{max})}{X_P P_{max}(1 - P_{ON})} \tag{25a}$$

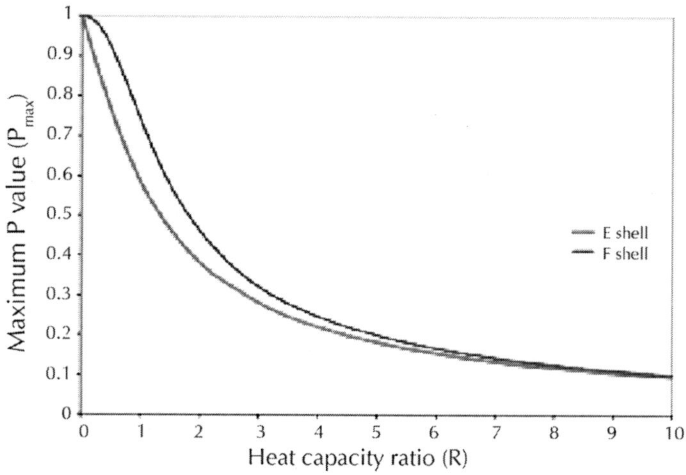

Figure 10: Comparison of maximum P value between E shell and F shell.

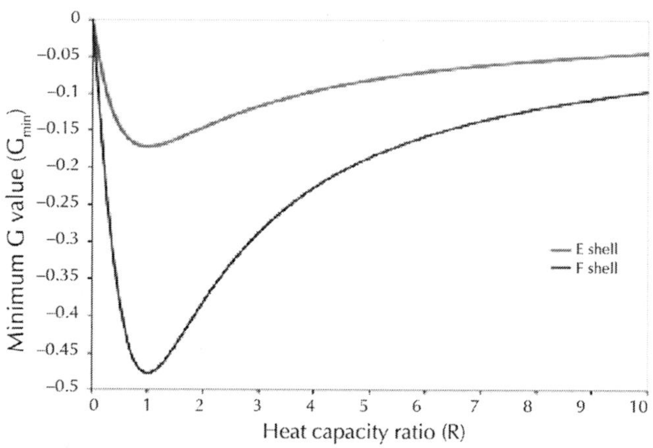

Figure 11: Comparison of minimum G value (G_{min}) between E shell and F shell.

Equations for Intermediate Temperatures

Once the number of heat exchangers and thereby P are fixed, temperatures of hot and cold fluids for all exchangers can be determined. From Eqs. (14), (17), (18) and (20), finite difference equation is formed for hot fluid temperature of nth exchanger and it is shown in Eq. (26).

$$(1 - P)T_{n+1} - (1 - PR)T_n = PRt_{N+1} - PT_1 \tag{26}$$

$$\text{Let } k = PRt_{N+1} - PT_1 \tag{27}$$

Here 'k' is constant since thermal effectiveness of individual shell (P), heat capacity ratio (R), inlet temperature of hot fluid (T_1) and exit temperature of cold fluid are known. The complete solution of Eq(26) contains two parts—complementary function and particular integral, and it is given in Eq. (28)

$$T_n = B\left[\frac{1 - PR}{1 - P}\right]^n + \frac{k}{PR - P} \tag{28}$$

where 'B' is arbitrary constant, which can be evaluated by applying the boundary condition: when n = 1,Tn = T_1

$$B = \left[\frac{1 - P}{1 - PR}\right]\left[T_1 - \frac{k}{PR - P}\right] \tag{29}$$

Substituting Eq. (29) in Eq. (28), Eq. (30) is obtained to calculate inlet temperature of hot fluid for nth exchanger. L'Hospital's rule helps to arrive at Tn, for R = 1.

$$\text{for } R \neq 1,\quad T_n = \left[T_1 - \frac{Rt_{N+1} - T_1}{R - 1}\right]\left[\frac{1 - PR}{1 - P}\right]^{n-1} + \frac{Rt_{N+1} - T_1}{R - 1} \tag{30}$$

$$\text{for R} = 1, \quad T_n = T_1 + \frac{P(n-1)(t_{N+1} - T_1)}{1-P} \tag{30a}$$

After calculating T_2, T_3, and T4 and so on, cold fluid temperature profile across all exchangers can be calculated using Eq. (31). Here, tn does not represent either cold fluid inlet or outlet temperature of nth exchanger as Tnrepresents for hot fluid. It represents cold fluid inlet temperature of $(N - n + 1)^{th}$ exchanger. For example, tNrepresent cold fluid inlet temperature of 1st exchanger.

$$t_n = \frac{(PR-1)T_{N+1-n} + T_{N+2-n}}{PR} \tag{31}$$

HANDLING ONE DESIGN PROBLEM

With all these relationships in hand, how does one solve a practical problem of designing multiple shell and tube heat exchangers? Let me illustrate this with an example, which will also help summarizing the essential results. Let us consider a heat exchanger design problem for a service shown in Table 1, where shell side is hot fluid cooled from 97.9 °C to −44.9 °C and tube side is cold fluid heated from −50.9 °C to 82.8 °C. In this case, temperature cross is too large [(82.8 − (−44.9) = 127.7 °C]. Here, design is established for both cases—E shell and F shell. The theoretically feasible maximum temperature cross (at $P = P_{max}$ or $G = G_{min}$) for single E shell and single F shell are 25.5 °C and 70.9 °C respectively. It means by providing infinite area cold fluid can be heated up to −19.4 °C using single E shell and up to 26 °C using single F shell, but never more than these values. Here our requirement is 82.8 °C. So it becomes inevitable to use multiple shells. The minimum required number of E shells and F shells that are theoretically feasible, are 9.9 and 4.95 which corresponds to ($X_P = 1$ or $P = P_{max}$ or $G = G_{min}$). It means minimum 10 E shells or 5 F shells with infinite area of individual exchanger, are required to meet the given terminal temperatures. This area can be reduced drastically by increasing the number of shells, on the other hand by decreasing X_P until F_T is more than critical value. For E shell case, theoretical number of E shells required is 13.33 for $X_P = 0.875$. It is rounded to 14 with $F_T = 0.8$. Of course $X_P = 0.9$ yields 12.57 exchangers, rounded to 13 with $F_T = 0.76$ which is conservative. Similarly for F shell, $X_P = 0.8$ requires 9.71 exchangers, rounded to 10 with $F_T = 0.91$ for each individual shell. When we connect only 9 exchangers, then F_T for each shell is 0.89. It is also acceptable since it is very close to 0.9. But let us continue with 10 exchangers for further analysis. P and F_T values of individual exchanger, for both cases are given in Table 2. One can select 7 number of F shells instead of 10 by adopting the fact that one 2–4 heat exchanger is equivalent to two 1–2 heat exchangers. Of course it could have been correct if we had determined number of E shells with $F_T \geq 0.9$. Two 1–2 heat exchangers can be considered equivalent to one 2–4 heat exchanger only when F_T of both 1–2 exchangers are equal and also equal to F_T of one 2–4 exchanger. If we select 7 number of F shells, then F_T of each exchanger becomes 0.8 which falls in the steep slope region as shown in Fig. 3. Obviously, it cannot be accepted since area tends to ∞ in this region. After number of exchangers for given duty and thereby P are fixed, temperature profiles for both hot and cold fluids across all heat exchangers can be determined from Eq. (30)and Eq. (31). Simulation results for E shell and F shell cases are presented in Table 3 and Table 4respectively. It can be seen from Table 3 that outlet temperatures of hot and cold fluids are almost same for all exchangers. If we had selected 13 exchangers instead of 14, some heat exchangers in this series would have handled small temperature crosses (may be approximately up to 2 °C). From Table 4, it can be seen that all F shell exchangers handle temperature crosses and it ranges from 2.53 °C (last exchanger) to 5.8 °C (first exchanger). Rigorous simulation to calculate EMTD for F shell should include correction factor proposed by Rozenman and Taborek (1971), but it is neglected in Table 4 calculation, assuming no thermal leakage. Fig. 12 and Fig. 13 show the temperature profiles of hot and cold fluid for E shell and F shell cases respectively. Heat duty presented in abscissa of Fig. 12 and

Fig. 13 are only representative of number of exchangers. Actual value can be estimated with the help of thermal data (specific heat vs temperature) and terminal temperatures of individual exchanger using Eq. (32).

$$Q_n = MC_{pn}(T_{n+1} - T_n) = U_n A_n (EMTD)_n \tag{32}$$

Table 1: Process parameters for a case study

Process parameters	Shell side	Tube side
Mass flow rate (kg/h)	264349	292055
Temperature (°C)	97.9/−44.9	−50.9/82.8
Density (kg/m³)	714.8/856	862/715
Viscosity (cP)	0.334/1.953	2.076/0.33
Specific heat (kJ/kg K)	3.47/2.28	2.27/3.25
Thermal conductivity (W/m K)	0.186/0.251	0.251/0.185
Fouling resistance (m² K/W)	0.000267	0.000267
Inlet pressure (kg/cm²) (a)	37.6	19.8
Allowable pressure drop (kg/cm²)	5	5

Table 2: Details of parameters used to determine number of E shells and F shells

Parameters	E shell	F shell
Shell side inlet temperature (T_{in})	97.9 °C	97.9 °C
Shell side outlet temperature (T_{out})	−44.9 °C	−44.9 °C
Tube side inlet temperature (t_{in})	−50.9 °C	−50.9 °C
Tube side outlet temperature (t_{out})	82.8 °C	82.8 °C
R	1.068	1.068
P overall	0.898	0.898
P_{max}	0.566	0.714
G_{min}	−0.1713	−0.4766
Minimum number of shells theoretically feasible	9.9	4.95

X_P	0.875	0.8
Number of shells (calculated)	13.33	9.71
Number of shells (selected)	14	10
X_P (improved)	0.855	0.79
P value of individual shell	0.484	0.564
Correction factor (F_T)	0.8	0.91

Table 3: Simulation results for E shell case: boundary temperatures and EMTD of each shell

Shell no.	T_{in} (°C)	T_{out} (°C)	t_{in} (°C)	t_{out} (°C)	LMTD (°C)	EMTD (°C)
1	97.9	82.78	68.65	82.8	14.61	11.71
2	82.78	68.63	55.4	68.65	13.68	10.97
3	68.63	55.38	42.99	55.4	12.8	10.27
4	55.38	42.98	31.38	42.98	11.99	9.61
5	42.98	31.36	20.5	31.38	11.23	9
6	31.36	20.49	10.32	20.5	10.5	8.43
7	20.49	10.3	0.79	10.32	9.84	7.89
8	10.3	0.78	−8.13	0.79	9.21	7.39
9	0.78	−8.14	−16.48	−8.13	8.62	6.91
10	−8.14	−16.49	−24.3	−16.48	8.07	6.47
11	−16.49	−24.31	−31.62	−24.3	7.56	6.06
12	−24.31	−31.63	−38.48	−31.62	7.08	5.67
13	−31.63	−38.48	−44.89	−38.48	6.62	5.31
14	−38.48	−44.9	−50.9	−44.89	6.2	4.97

Table 4: Simulation results for F shell case: boundary temperatures and EMTD of each shell

Shell no.	T_{in} (°C)	T_{out} (°C)	t_{in} (°C)	t_{out} (°C)	LMTD (°C)	EMTD (°C)
1	97.9	77	63.24	82.8	14.42	13.15
2	77	57.96	45.4	63.24	13.15	11.99
3	57.96	40.59	29.14	45.4	11.99	10.94

4	40.59	24.75	14.31	29.14	10.94	9.97
5	24.75	10.31	0.79	14.31	9.97	9.09
6	10.31	−2.86	−11.53	0.79	9.09	8.29
7	−2.86	−14.86	−22.78	−11.53	8.29	7.56
8	−14.86	−25.81	−33.03	−22.78	7.56	6.89
9	−25.81	−35.79	−42.38	−33.03	6.89	6.29
10	−35.79	−44.9	−50.9	−42.38	6.29	5.73

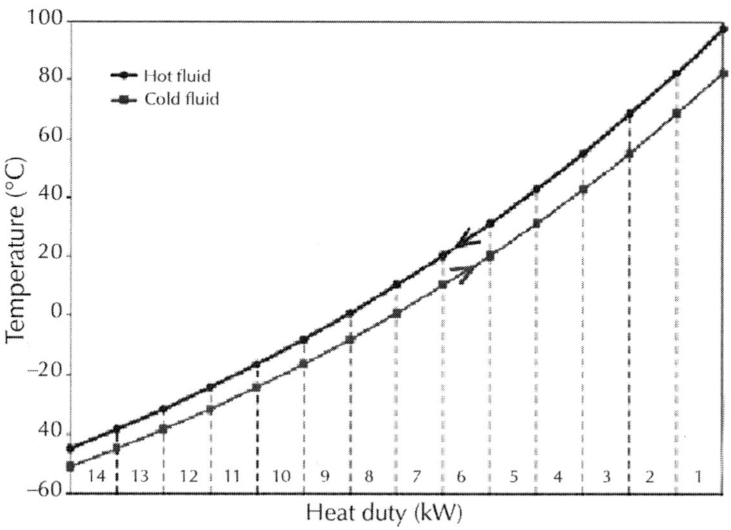

Figure 12: Hot and cold fluid temperature profiles across exchangers for E shells connected in series case (shell numbers are given at the bottom).

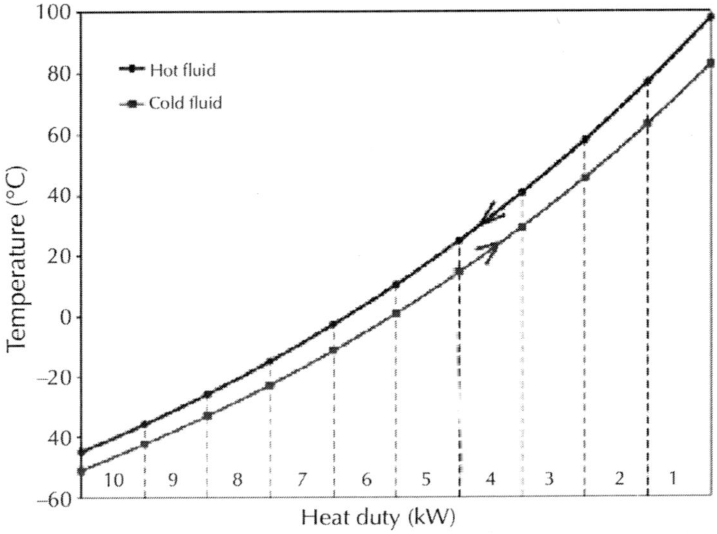

Figure 13: Hot and cold fluid temperature profiles across exchangers for F shells connected in series case (shell numbers are given at the bottom).

Heat duty may vary across exchangers, though they are geometrically identical. It is due to the fact that temperature differences of stream in each exchanger need not to be identical and also due to variation in specific heat. If we assume specific heat varies linearly for Table 1 service, then individual heat exchanger duty can be estimated through Eq. (32). Now heat duty and EMTD are fixed for all exchangers. To calculate overall heat transfer coefficient of individual exchanger, it is essential to know heat transfer area (A). For this, heat exchanger geometry is assumed (shell diameter, tube diameter, tube length, tube layout, baffle spacing, baffle cut, etc.) and A can be calculated. Required overall heat transfer coefficient (U_{req}) for assumed heat exchanger geometry can be calculated using Eq. (32) since Q and EMTD are already known. Selection of heat exchanger geometry is usually trial and error basis, so that overall heat transfer coefficient estimated based on standard correlations available in literature, i.e., actual heat transfer coefficient of chosen geometry (U_{act}) exceeds U_{req} and till desired % overdesign ($U_{act} \times 100/U_{req}$) is achieved. Details of correlations for estimating shell side heat transfer coefficient – Bell Delaware method (Taborek, 1988), stream analysis method – and tube side heat transfer coefficient can be found in Serth, 2007. It requires hot and cold fluid flow rates and heat release curves (hydraulic and thermal properties variation with temperature). U_{req} and U_{act} and hence % overdesign vary across exchangers and it is due to variation of EMTD, hydraulic and thermal properties change over exchangers. Different heat exchanger geometries were tried using HTRI Inc. software and the final design which satisfies process requirements, TEMA standards—geometry constraints, requirements to avoid ρV^2 and vibration problems and stream analysis are shown inTable 5. In order to get desired tube side velocity, four tube passes was considered in E shell case. As stated earlier, F_T is almost same for both 1–2 and 1–4 heat exchangers. Table 6 and Table 7 show the U_{req} in each exchanger and it varies from 810.15 to 534.18 W/m^2 K for E shell case and from 898.94 to 591.37 W/m^2 K for F shell case respectively. Arithmetic mean of EMTD, ($\overline{EMTD}$) and arithmetic mean of$U_{req,}$ ($\overline{U_{req}}$) for all heat exchangers in series can be considered as a representative value for overall unit. These values $\overline{EMTD}$ and $\overline{U_{req}}$ are 7.9 °C and 671.66 W/m^2 K for E shell case and 9 °C and 744.62 W/m^2 K for F shell case. Similar to$\overline{U_{req}}$, $\overline{U_{act}}$ can be estimated by taking arithmetic mean of individual exchangers. The estimated $\overline{U_{act}}$ by HTRI Inc., are 694 and 772 W/m^2 K for E shell and F shell cases. For the purpose of capital cost comparison and the heat exchanger geometry were selected so that overall % over design ($\overline{U_{act}} \times 100 / \overline{U_{req}}$) is almost same for both cases.

Table 5: Details of heat exchanger geometry and performance parameters for the case study

Parameters	**E shell**	**F shell**
Number of shells in series	14	10

Number of shell passes	1	2
Number of tube passes	4	4
Total heat transfer area (m^2)	406.1 × 14 = 5685.4	450.5 × 10 = 4505
Shell ID (mm)	1150	1100
Tube OD × thickness × length (mm)	19.05 × 2 × 4600	19.05 × 2 × 5600
Tube pitch (mm)	25.4, triangular (30°)	25.4, triangular (30°)
Shell side pressure drop (kg/cm^2)	3.73	4.566
Tube side pressure drop (kg/cm^2)	4.8	4.874
Design pressure of shell side/tube side (kg/cm^2) (a)	40.8/21.1	40.8/21.1
Baffle type	Double segmental, vertical	Double segmental, vertical
Baffle cut (% diameter)	25	26
Baffle spacing (mm)	239	573
Number of baffles/shell	13	8
Number of tubes/shell	1520	1376
Shell side cross flow velocity (m/s)	0.54	0.67
Shell side velocity thorough baffle window (m/s)	0.42	1.03
Tube side velocity (m/s)	1.55	1.72
Overall heat transfer coefficient $\overline{U}_{act}$ (W/m^2 K)	694	772
Shell side film resistance (%)	31.37	26.43
Shell side fouling resistance (%)	13.88	15.45
Tube wall resistance (%)	4	4.45
Tube side fouling resistance (%)	17.82	19.84
Tube side film resistance (%)	32.93	33.83
Stream fraction (cross flow)	0.67	0.78
Stream fraction through baffle hole-tube OD clearance	0.17	0.08
Stream fraction through shell ID-baffle OD clearance	0.12	0.07
Stream fraction through shell ID-bundle clearance	0.05	0.06

Table 6: Required heat transfer rate and overall heat transfer coefficient in each exchanger for E shell case: shell side flow rate = 264,349 kg/h, A = 406.1 m^2

Shell no.	T_{in} (°C)	T_{out} (°C)	EMTD (°C)	C_p (kJ/kg K)	Q (kW)	U_{req} (W/m^2 K)
1	97.9	82.78	11.71	3.47	3852.62	810.15
2	82.78	68.63	10.97	3.379	3510.91	788.10
3	68.63	55.38	10.27	3.288	3199.06	767.04
4	55.38	42.98	9.61	3.197	2910.98	745.90
5	42.98	31.36	9	3.106	2650.23	725.12
6	31.36	20.49	8.43	3.015	2406.53	702.96
7	20.49	10.3	7.89	2.924	2187.90	682.84
8	10.3	0.78	7.39	2.833	1980.43	659.90
9	0.78	−8.14	6.91	2.742	1796.00	640.02
10	−8.14	−16.49	6.47	2.651	1625.44	618.63
11	−16.49	−24.31	6.06	2.56	1470.02	597.33
12	−24.31	−31.63	5.67	2.469	1327.11	576.36
13	−31.63	−38.48	5.31	2.378	1196.13	554.69
14	−38.48	−44.9	4.97	2.287	1078.14	534.18
			$\overline{EMTD} = 7.9$			$\overline{U_{req}} = 671.66$

Table 7: Required heat transfer rate and overall heat transfer coefficient in each exchanger for F shell case: shell side flow rate = 264,349 kg/h, A = 450.5 m^2

Shell no.	T_{in} (°C)	T_{out} (°C)	EMTD (°C)	C_p (kJ/kg K)	Q (kW)	U_{req} (W/m^2 K)
1	97.9	77	13.15	3.47	5325.38	898.94
2	77	57.96	11.99	3.338	4666.90	864.00
3	57.96	40.59	10.94	3.206	4089.20	829.71
4	40.59	24.75	9.97	3.074	3575.48	796.06
5	24.75	10.31	9.09	2.942	3119.50	761.77
6	10.31	−2.86	8.29	2.81	2717.49	727.64
7	−2.86	−14.86	7.56	2.678	2359.76	692.87
8	−14.86	−25.81	6.89	2.546	2047.14	659.53
9	−25.81	−35.79	6.29	2.414	1769.06	624.31
10	−35.79	−44.9	5.73	2.282	1526.54	591.37
			$\overline{EMTD} = .9$			$\overline{U_{req}} = 744.62$

COST ESTIMATION AND OPTIMIZATION

Several methods – six-tenth method, Guthrie method, Pikulik method, Corripio method, Purohit method, Hall method, Vatavuk method, and Matches Co. method – for capital cost estimation are available in literature and those are summarized in Taal et al. (2003). In this work, Purohit method is used to estimate the cost since it takes into account many input parameters: front, shell, and rear TEMA types, tube/shell/channel and tube sheet material, heat transfer area, shell ID, tube length, number of tube passes, tube OD, tube pitch, tube layout angle, type of expansion joint, shell side and tube side design pressure, tube material gauge (BWG). In this method, cost of the baseline exchanger (E type shell with welded carbon steel, 14 BWG tube, tube passes 1 or 2, tube length 6 m, shell side and tube side design pressure ≤10.5 kg/cm², Material of construction of all parts—carbon steel) is given in Eq. (33).

$$b_c \left[\frac{6.6}{1 - e^{[(7-D_i)/27]}} \right] pfr \tag{33}$$

Here, D_i = shell ID or bundle diameter of a kettle reboiler; p = cost multiplier for tube OD, pitch and layout angle, f = cost multiplier for TEMA-type front head, r = cost multiplier for TEMA-type rear head.

After the exchanger base price is estimated, the free-on-board (f.o.b.) January 1982 price of any shell and tube heat exchanger can be estimated in Eq. (34).

$$E_b = [b_c\, 1 + C_T\, A]N \tag{34}$$

Here A = heat transfer area; N = number of exchangers; C_T = sum of cost corrections of base price cost, which include corrections for shell type, front head and rear head types, tube length, number of tube passes, tube gauge, shell side and tube side design pressures, materials of construction. Details of calculation procedures for C_T can be found in reference Purohit (1983). An exchanger price estimated by Purohit method can be escalated to a price for a date beyond January 1982 by means of an escalation index. One such index that applies to process heat exchangers is the Fabricated Equipment component of the Chemical Engineering Plant Cost Index. Other possibilities are the Marshall & Swift Equipment Cost Index, the Nelson Refinery Cost Index and the U.S. Bureau of Labor Statistical General purpose Machinery and Equipment Cost Index. Cost multipliers and corrections for the Table 5 design is shown inTable 8. The cost estimated based on Purohit method for E shell case shows 27.4% more than F shell case. If piping cost is included, it will be still higher.

Table 8: Parameters required for cost estimation and comparison of cost for Table 5 design

	E shell	**F shell**
p	0.85	0.85
f	1	2
r	0.9	0.9
b ($/ft^2)	6.66	6.83
Cost correction for shell type	0	0.2
Cost correction for tube length	0.3	0.058
Cost correction for number of tube passes	0.03	0.03
Cost correction for shell side design pressure	0.34	0.34
Cost correction for tube side design pressure	0.05	0.05
Cost corrections of other parameters	0	0
Total cost correction (C_T)	0.725	0.66
Exchanger price on January 1982 ($)	703598.2/14 shells	552357.9/10 shells
Chemical Engineering Plant Cost Index on January 1982	324.5	324.5
Chemical Engineering Plant Cost Index on January 2009	603.4	603.4

Exchanger price on January 2009 ($)	1308324.0/14 shells	1027096.0/10 shells

An attempt could be made to decrease the total cost shown in Table 8, by decreasing the size of each exchanger but with increased number of exchangers so that total heat transfer area is constant. According toHall et al. (1990) capital cost of N exchangers with total area (A_T) can be estimated using Eq. (35).

$$C = N\left[a + b\left(\frac{A_T}{N}\right)^c\right] \tag{35}$$

where a, b, and c are cost law coefficients which vary according to materials of construction, pressure rating and type of exchanger. Values for c are typically between 0.6 and 0.9 (Smith, 2005), which reflect the economies of scale generally observed in chemical process plant. If cost law coefficients of particular heat exchanger and A_T are known, one can estimate optimum number of shells with F_T constraint, to minimize C. An additional constraint of individual exchanger area has to be considered from the limitation imposed by the plant owner. Generally, heat exchangers with floating head rear end configurations or U-tube, which have removable tube bundles, are limited in size by crane handling capacities and therefore weight is limited to 10 tons which corresponds to 450–500 m^2 of heat transfer area. Some larger plants permit tube bundles up to 15–20 tons. For fixed tube sheet heat exchangers, very large heat transfer area (2000 m^2 or more) can be accommodated since bundle weight has no limitation but considering fabrication capabilities, transportation facilities and plot area limitations.

CONCLUDING REMARKS

Based on equations developed in this paper, one can determine number of shells required and temperature profile across all exchangers for both E shell and F shell cases. Though Eq. (25) and

Eq. (25a) are common, the parameters – P_{max}, G_{min}, X_P and thereby P – are different for E shell and F shell. The X_P value for F shell may be chosen as 0.85 for $R \leq 0.6$, 0.8 for $(0.6 < R \leq 2)$, and 0.85 for $R > 2$, to avoid steep slope region in $F_T$chart. In general, $F_T \geq (0.75$ or $0.8)$ for E shell and $F_T \geq 0.9$ for F shell can be considered to stay away from steep slope region. Since one F shell is equivalent to 2 E shells, required number of F shells may also be determined by using E shell equations and keeping X_p in the range 0.75–0.8 to ensure $F_T \geq 0.9$. After determining P value of individual exchanger, temperature profiles for hot and cold fluids across all the exchangers can be obtained with the help of Eq. (30) and Eq. (31). This is essential for further design. Though, F shells are associated with problems of leakage – thermal and physical, these offers certain advantages over E shells – less number of shells, thereby leading to savings in exchanger and piping cost, better shell side velocity resulting in lower fouling, higher heat transfer coefficient and lower operating cost. Thermal leakage can be overcome by providing an insulated longitudinal baffle. Physical leakage can be reduced by limiting the shell side pressure drop below 0.5 kg/cm^2 (some licensors prefer 0.35 kg/cm^2) and sealing the gap between longitudinal baffle and shell wall by using flexible strips. These seals may get damaged during tube bundle removal and consequently leakage deteriorates temperature profile. During the design phase particularly for large temperature cross, designs have to be carried out for both cases. It is necessary to take care of leakage problem in F shell heat exchanger is to be chosen.

REFERENCES

1. Ahmad, S., Linhoff, B. and Smith, R., 1988, Design of multipass exchangers; an alternative approach. ASME J Heat Trans, 110: 304–309.
2. Gulyani, B.B., 2000, estimating number of shells in shell and tube heat exchangers: a new approach based on temperature cross. ASME J Heat Trans, 122: 566–571.

3. Hall, S.G., Ahmad, S. and Smith, R., 1990, Capital cost targets for heat exchangers networks comprising mixed materials of construction, pressure ratings and exchanger types. Comp Chem Eng, 14: 319–335.
4. Kern, D.Q., (1997). Process Heat Transfer. (McGraw-Hill, USA), pp. 144–177
5. Moita, R.D., Fernandes, C., Matos, H.A. and Nunes, C.P., 2004, A cost-based strategy to design multiple shell and tube heat exchangers. ASME J Heat Trans, 126: 119–130.
6. Mukherjee, R., 1998, Broaden your heat exchanger design skills. Chem Eng Prog, 35–43.
7. Nagle, W.M., 1933, Mean temperature differences in multi pass heat exchangers. Ind Eng Chem, 25: 604–609.
8. Perry, R.H. and Green, D.W., (1997). Perry's Chemical Engineers Handbook: Section 11, Heat Transfer Equipment (7th Ed.). (McGraw-Hill, New York), pp. 33–35
9. Ponce-Ortega, J.M., Gonzaletz, S. and Gutierraz, J., 2008, Design and optimization of multipass heat exchangers. Chem Eng Proc, 47: 906–913.
10. Purohit, G.P., 1983, estimating costs of shell-and-tube heat exchangers. Chem Eng, 22: 56–67.
11. Rice, R.G. and Do, D.D., (1995). Applied Mathematics and Modeling for Chemical Engineers. (John Wiley & Sons, New York), pp. 164–183
12. Rozenman, T. and Taborek, J., 1971, the effect of leakage throughout the longitudinal baffle on performance of the two-pass shell exchanger. AIChE Symp Series, 118(68): 12–20.
13. Seider, W.D., Seader, J.D. and Lewin, D.R., (2004). Product & Process Design Principles, Synthesis, Analysis, and Evaluation (2nd Ed.). (John Wiley & Sons, New York), p. 426
14. Serth, R.W., (2007). Process Heat Transfer: Principles and Applications. (Elsevier Science and Technology Publications), pp. 245–326
15. Shenoy, U.V., (1995). Heat Exchanger Network Synthesis—

Process Optimization by Energy and Resources Analysis. (Gulf Publishing Company, Houston). Chapter 6. pp. 255–264

16. Smith, R., (2005). Chemical Process Design and Integration. (John Wiley & Sons, England), pp. 324–329
17. Taal, M., Bulatov, I., Klemes, J. and Stehlik, P., 2003, Cost estimation and energy price forecasts for economic evaluation of retrofit projects. Appl Therm Eng, 23: 1819–1835.
18. Taborek, J., (1988). Shell and Tube Heat Exchangers: Heat Exchanger Design Handbook (Hemi Sphere Publishing Corp, New York).
19. Underwood, A.J.V., 1934, the calculation of the mean temperature difference in multi pass heat exchangers. J Inst Petroleum Tech, 20: 145–158.
20. Wales, R.E., 1981, Mean temperature difference in heat exchangers. Chem Eng, 88: 77–81.

Citations

CHAPTER 1

Katarzyna St pczy ska, Henryk Łukowicz, and Sławomir Dykas, Diverse Configurations of the Boiler Feed Pump Drive for the Ultra-supercritical 900-MW Steam Plant, doi:10.1186/2251-6832-3-3.

CHAPTER 2

Anderson, K., Devost, M., Pakdee, W. and Krishnamoorthy, N. (2013) STAR CCM+ CFD Simulations of Enhanced Heat Transfer in High-Power Density Electronics Using Forced Air Heat Exchanger and Pumped Fluid Loop Cold Plate Fabricated from High Thermal

Conductivity Materials. Journal of Electronics Cooling and Thermal Control, 3, 144-154. doi: 10.4236/jectc.2013.34016.

CHAPTER 3

O. Prakash, D. Kumar, and Y. Dwivedi, "Effects of Thermal Diffusion and Chemical Reaction on MHD Flow of Dusty Visco-Elastic (Walter's Liquid Model-B) Fluid," Journal of Electromagnetic Analysis and Applications, Vol. 2 No. 10, 2010, pp. 581-587. doi: 10.4236/jemaa.2010.210075.

CHAPTER 4

Hasan, M., Rageb, A. and Yaghoubi, M. (2012) Investigation of a Counter Flow Microchannel Heat Exchanger Performance with Using Nanofluid as a Coolant. Journal of Electronics Cooling and Thermal Control, 2, 35-43. doi: 10.4236/jectc.2012.23004.

CHAPTER 5

Naili, N. , Attar, I. , Hazami, M. and Farhat, A. (2012) Experimental Analysis of Horizontal Ground Heat Exchanger for Northern Tunisia. Journal of Electronics Cooling and Thermal Control, 2, 44-51. doi:10.4236/jectc.2012.23005.

CHAPTER 6

Toledo-Velázquez, M. , Quinto-Diez, P. , Alzelmetti-Zaragoza, J. , Galvan, S. , Abugaber-Francis, J. and Reyes-León, A. (2014) Delaware Method Improvement for the Shell and Tubes Heat Exchanger Design. Engineering, 6, 193-201. doi: 10.4236/eng.2014.64023.

CHAPTER 7

Zhanying Zheng, Anthony M. Johnston, David F. Fletcher, Brian S. Haynes, Heat exchanger specification: Coupling design and surface performance evaluation, Chemical Engineering Research and Design, Volume 93, January 2015, Pages 392-401, ISSN 0263-8762, http://dx.doi.org/10.1016/j.cherd.2014.04.015.

CHAPTER 8

Raquel Y. Miura, Flavio C.C. Galeazzo, Carmen C. Tadini, and Jorge A.W. Gut, The Effect of Flow Arrangement on the Pressure Drop of Plate Heat Exchangers, doi: 10.1016/j.ces.2008.07.029.

CHAPTER 9

U. Vengateson, Design of multiple shell and tube heat exchangers in series: E shell and F shell, Chemical Engineering Research and Design, Volume 88, Issues 5–6, May–June 2010, Pages 725-736, ISSN 0263-8762, http://dx.doi.org/10.1016/j.cherd.2009.10.005.

Index

G

H

L

M

P

S

T